Introduction to Laser Physics with Applications in Fiber Optics

Steve Lee

EpiphanyBySteveLee.com

Introduction to Laser Physics with Applications in Fiber Optics

More information is available at: EpiphanyBySteveLee.com

Preface:

Most of the following information was compiled over the years for use as support material in teaching a number of informal classes on laser theory to undergrads, technicians, and engineers. This information is offered here in the hope that those who have an interest in lasers would find this unique treatment of the topic helpful in broaching the subject.

The primary goal in structuring this material as it is, was to make the mechanisms behind lasing accessible to as wide an audience as possible, ranging from those with limited or no prior exposure to Quantum Physics, to those with some grasp of the subject. To that end, we have made a concerted effort to *defang* the math in the initial sections of each chapter as much as possible, offering models and analogous examples that help demonstrate the essential concepts instead. These fundamental models include the description of the general function of the laser in terms of the basic "Heat Engine", as well as using the classical "ball and spring" model of the atom as an introductory first approximation of atomic systems, etc.

Though these more classical models are conceptually useful in introducing the topics, we will find that they cannot explain certain key aspects of lasing (e.g. the "Metastable State"), requiring us to refine our picture by adopting the Quantum approach to our model. Consequently, as we progress we will begin to gradually introduce these more refined concepts, while still doing our best to keep the initial exposure to each as non-life threatening as possible.

To that end, we have tried to keep the ugliest math to a minimum, holding the more rigorous treatment of each subject for the latter sections. For those topics which require more depth for a full understanding of lasing (e.g. the Dipole Selection Rule), we placed the more rigorous mathematical treatments in the appendices, as a way to provide those who seek a deeper level of understanding to the mechanisms of lasing what they need as well.

Hopefully most readers will find this more "user friendly" approach helpful, allowing each individual to explore the various aspects of laser theory to whatever depth suits them. Having made this effort, we therefore strongly encourage even the casual reader to resist any urge towards the "fight or flight" impulse to bail at the first sight of an equation, hoping they will rise to the challenge by taking advantage of the conceptual discussion offered, while developing sufficient nerve to stick with it to the bitter end.

And should you find this information (and/or the accompanying attempt at lighthearted chaff) great conversational fodder at parties, we hope you will take the time to check out some of the other works on our web page as well, at "*EpiphanyBySteveLee.com*".

Steve Lee

Steve Lee

"It has become appallingly obvious that our technology has exceeded our humanity."

– Albert Einstein

Table of Contents:

Chapter 1: Introduction:

Since the emergence of working laser from the realm of Science Fiction lore in the late 1950's, the laser has found its way into an extremely wide range of modern applications. These include everything from highly delicate eye and heart surgery, to the more brute force setting of assembly line fabrication (e.g. of metal, wood and plastics), as well as virtually every conceivable application in between.

Some of the more imaginative applications of lasers include CAD prototype design and fabrication using computer controlled laser sources to raster through and thus catalyze a light sensitive liquid polymer to instantly construct virtually any conceivably shaped plastic prototype on demand; fiber optic telecommunication links, achieving ultra high bandwidth with data throughputs on the order of 10 Gigabits per second (easily increasing the information exchange rate throughout the world literally *hundreds of thousands* of times greater than it was just three decades ago); ultra high precision geographical monitoring and surveying, capable of detecting physical feature changes within pico-meters (10^{-12} meters); security and surveillance devices capable of measuring micro-miniature vibrations induced on walls and windows due to footsteps and voice conversations, allowing law enforcement groups (and others) to ease-drop on conversations from over 100 yards away simply by bouncing a laser beam off a window and monitoring the voice induced vibrations reflected off the glass; etc.

Since the fundamental operation of a laser depends on a variety of physical mechanisms, any comprehensive description of the more basic principles of lasers will require the use of several different disciplines in physics, including Quantum physics and Electrodynamics. Though we will avoid most of the more painful and involved derivations as much as possible through the main body of this text, in order to discuss a topic as complex as this, some use of these disciplines will be required since we will find that the basic classical models fail to accurately explain many of the key aspects of lasing. This is particularly true for those aspects centering around atomic and molecular behaviors, since objects the size of atoms are far too small to be physically observed and hence accurately analyzed in any other way.

To begin our discussion, we first describe the laser as an example of a Carnot or "Heat" Engine, underscoring the operation of a laser as a refining mechanism that converts raw energy into a highly specialized output (coherent monochromatic light). In describing the concept of a Carnot engine, we mention the familiar internal combustion engine in passing as an analogue for the laser as a refining "engine", etc. As we proceed, we will make frequent use of such analogies, in an attempt to use common objects to help explore less familiar concepts at the heart of how the laser does what it does. We will note that in general, the more refined and sophisticated the output of any such "engine", the more complex the mechanism employed to do that refinement.

In the case of lasers, an enormous number of distinct individual "machines" (typically $>10^{20}$) are incorporated into the fundamental heart of the device (the "Gain Media"). Fortunately for us, nature has generously provided most of the essential mechanisms needed to enable that many individual elements to work in unison in order to produce the refined output we seek. What's more, despite such a vast number of individual atoms or molecules involved, we find that the overwhelming majority of them operate with extreme precision and predictability, not to mention a nearly *zero* defect rate!

Once we have discussed the basic role of the gain media (the specifically chosen atoms or molecules responsible for storing and "tuning" the energy emitted by the laser), we then introduce the classical "ball and spring" model of the atom, which we will use to help understand how atoms and molecules extract specific wavelengths of light from the raw energy injected into the gain media. This "ball and spring" model will help us understand the effects a material has on light as a function of "resonance", "absorption", "chromatic dispersion", etc., giving us a basic picture of how the atom works. As useful as this model is as a basic "first approximation" however, we will find that such classical descriptions fail to explain other key characteristic behaviors of atoms and molecules (e.g. the storage of energy in the "Metastable State", etc.).

To fully understand these behaviors, we will need to introduce concepts and models developed using the Quantum picture of the atom. We therefore make the transition from the more classical picture of things and move into the topic of the Bohr model of the Atom and the Quantum view of the Simple Harmonic Oscillator (SHO). Using these models we introduce several common gas lasers (e.g. HeNe and CO_2) which we will find serve as a good introduction for the basic lasing model. Building on these concepts, we then transition into a basic discussion of semiconductors, leading to a brief overview of several now popular semiconductor laser diodes, discussing both their benefits as well as some of their short comings.

Next we shift gears slightly to discuss a specific application of lasers in telecommunications, including an overview of many key components used in Fiber Optic telecommunications (the glass waveguide, optical grating filters, Optical Time Domain Reflectometers, etc.). Finally, we conclude with several more advanced topics in the appendices for completeness, including a more in-depth discussion on Quantum topics (e.g. Hydrogen atom state functions, an application of Schrödinger's Equation to molecular vibrational modes, the Dipole Selection rule, and a brief survey of Dirac Notation used in Quantum Physics, etc.).

Though many of these theoretical descriptions sometimes seem "awkward" in terms of what our "common sense" might lead us to expect (particularly some of the concepts developed in Quantum Physics), the proof of their validity is found in the *mountains* of hard data generated through more than a century of experimentation. This of course includes data related to all of what was at first considered "anomalous" spectra given off by various gas lamps (e.g. hydrogen, methane, neon, argon, etc.). Historically, it was the exploration of these rather "odd" spectra that led (in part) to the development of Quantum Physics during the early part of the twentieth century, and are essential to a solid understanding of how the most fundamental parts of a laser work, being impossible to explain using the models developed through any other means.

"Carnot Engines"

We begin our discussion by introducing what is known as a "Carnot engine". This term was coined back in the early 1800's by the likes of Nicolas Carnot who used it to describe any mechanism (i.e. "engine") powered by some external "raw" energy source (e.g. fire) that drives its internal components in order to produce some *refined*, specialized output. In general terms, this refinement process is accomplished by exploiting the basic thermodynamic properties of molecules/gases in order to drive a mechanism that performs some specialized task (e.g. grind wheat, pump water, turn the wheels of the family car, etc.).

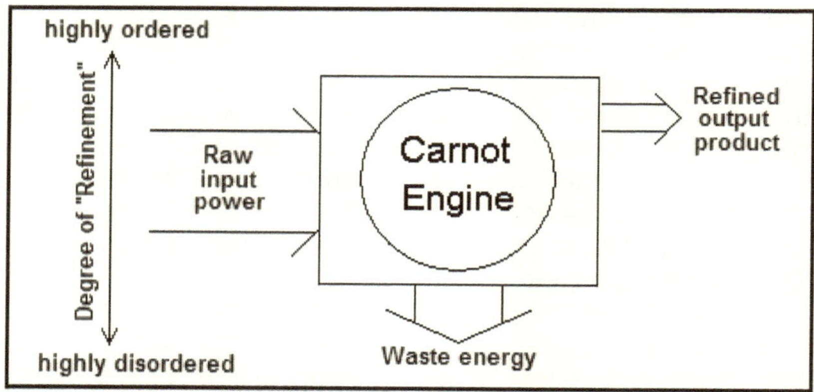

Figure 1.1 Diagram of the basic Carnot engine

The most common example of a Carnot engine of that period was the "Iron Horse", which was powered by nothing more than a wood or coal fire used to turn water into steam. The pressure from the energetic steam was then used to drive pistons, which were connected to a simple drive mechanism to focus the engine's raw mechanical energy to perform some useful work, such as turning the main wheels of the train's locomotive, or powering a weaver's loom, spinning a saw blade, etc.

One of the more fundamental aspects of any Carnot engine, is that they all take some amount of "raw energy" to drive their mechanism in order to produce a specialized output. This resulting refined output typically represents only a small fraction of the original input power, implying that a large portion of the energy we inject into all such engines (including the internal combustion engine) is wasted in the process.

Since the Law of Conservation of Energy tells us that energy is never created nor destroyed, we know that the *difference* between the total energy we inject into a system, and the small percentage of energy we find in the refined output has to go somewhere. Therefore we conclude that the process of refining the raw input energy into the precise and highly structured output effectively *costs* us some amount of the total energy injected into the engine.

Drawing upon the average family car as an example (something we note is itself based on technologies which date back several centuries or more[1]), we find that for every $1.00 of gasoline we put into the machine, only about *12 cents* of that fuel's energy actually goes into turning the wheels of the car (i.e. making the average car only about 12% efficient). The rest of the input energy is wasted either as heat (e.g. friction in the brake pads during our stop and go "rush hour" driving, heat pushed out the radiator in an attempt to keep the engine from over-heating, etc.), or as energy used to position the various components within the engine (e.g. valves, camshafts, pistons, gears, etc.) – all as an unavoidable part of the operation of the engine that enables it to convert part of the raw energy we feed it into a useful refined output.

---

1 For example, the spark plug can be traced back to a device known as Volta's pistol (ca. 1777), which was used to test for "mal air" ("bad air") in and around locations known to be associated with "*malaria*", i.e. swamps. Samples of swamp air were drawn into the iron chamber, wherein was a brass fitting in a glass tube which arced when connected to a charge source. If the air sample was volatile (e.g. methane) the arc ignited a mild explosion which could be used to eject a small lead ball.

In its general form, a Carnot engine (also known as a "Heat engine") has two "reservoirs": a "Hot" reservoir and a "Cold" reservoir. Molecules in the Hot reservoir can be excited by indirect flame (e.g. energetic steam produced in the boiling chamber), or through the direct expansion of an exploding fuel (e.g. the hot gases released in the pistons by igniting gasoline inside the piston cylinder). Consequently, these highly energetic molecules then exert a great deal of pressure on the walls of the container (e.g. engine cylinder), which we then use to push the piston, and thus turn the wheels of the vehicle (or spin a generator, etc.). In essence, *we allow the molecules in the hot gas to do what they naturally want to do, and in the process we have them do a little work for us along the way.*

Although it may seem a little odd at first blush to be drawing upon a steam or automobile engine during a discussion about the operation of a laser, we find that these examples offer us a crude, yet tangible analogy with which most people are at least superficially familiar. In order to convert the raw energy stored in the fuel molecule (e.g. methane: CH_4, propane: C_3H_8, ethanol: C_2H_5OH, or isooctane: C_8H_{18}) into the refined energy needed to power our machines, we must create a very complex mechanism specifically designed to harness and refine that energy, complete with a variety of precisely constructed and coordinated components. Each of these components perform specific tasks at precise intervals, in order to meet the overall objective of bringing order to the chaos which is the rapidly expanding gas of the ignited fuel.

To achieve this orchestrated precision in the internal combustion engine, we must literally *program* each event in the process by shaping or positioning the various mechanical components inside the engine (e.g. gears, lever arms, cam shafts, etc.). For example, the timing of the valve opening and closing at the top of the combustion chamber is controlled by small camshafts which rotate in-sync with the piston movements in such a way that the valves are pushed open at the *precise* instant they need to be open (the intake and exhaust phase), and closed when they need to be closed (the compression and ignition phase). Inadvertently opening or closing at the wrong moment could easily result in the valves being crushed by the piston, destroying the operation of the engine in the process.

Once people found they could coordinate and program such multiple split-second events with extreme mechanical precision using pre-shaped gears, cam shafts, etc., several clever inventors found they could then extend this technique to build engines capable of performing highly refined logical and mathematical operations, creating what became known as "*Analytic Engines*". The most prominent of these inventors[2] included Charles Babbage (ca. 1823) and Lord Kelvin (ca. 1875), who took a basic Carnot engine and used it to drive gears, rods and pulleys in such a way as to perform a sequence of highly specific mathematical operations, including addition, subtraction, multiplication and division. They then combined these various basic mathematical elements in such a way as to create calculating machines capable of accurately computing insurance actuary tables, predict the periodic ebb and flow of tide waters in a specific bay or inlet, etc.

In the process of refining this form of the Carnot engine, Babbage and Kelvin created the basic physical and theoretical building blocks that became the foundation upon which our modern computer systems are built (i.e. an input/output device, memory, the core processing engine, and a prescribed sequence of actions – i.e. the "program"). Taking this one step further, someone then combined the core processing ability of these "Analytic Engines" with the variable punch card technique used to change the pattern created by weaving looms, allowing these

---

2 It is interesting to note that Babbage was actually not the first to create such an analog computer. In fact, the technology to build such a device existed at least 2000 years ago, as clearly shown by the finding of what has become known as the "Antikythera Mechanism", found off the Greek island of Antikythera in 1901 in the remains of an ancient shipwreck. This device used carefully designed brass planetary gears (similar to our mechanical watches and clocks), to *program* the device with some very sophisticated knowledge about celestial mechanics, and served to accurately predict solar and lunar eclipses, the position of the sun, moon and several other celestial objects, etc. From inscriptions on the gears it is clear that it was constructed by at least 80 BC, if not earlier [1].

"Analytic Machines" to go from running only a single "hard wired" "list" of actions, into a device with variable programming, enabling these computing engines to perform a much wider range of tasks than the single task previously built into them via a cam shaft or gearing program. This technique effectively opened the door to the "modern" punch-card computer circa 1940. And the rest is, as they say, *history*.

Whether we are discussing a steam engine, the engine in our family car, an analytic engine, or a laser "engine", in the more general sense, all of these examples of the Carnot engine share at least two fundamental characteristics:

1) In order to do something as simple as convert one form of raw energy (e.g. chemical or mechanical) into a more refined and highly structured output, a complex mechanism is needed to perform this refinement process.

2) Any such conversion/refinement process *costs* us both *energy*, and *complexity*. In other words, *all* refining engines involve some complex processing elements, and *all* such processing mechanisms inevitably *use* and *waste* some amount of the energy injected into them to drive their refinement mechanisms.

As it turns out, this latter point is an unavoidable part of the "cost of doing business" in the universe. All such Carnot engines (whether they be man made machines, the mechanisms in our bodies[3], etc.) operate at less than 100% efficiency – i.e. converting a small portion of the raw energy injected into them into their refined output products.

Since a laser in a general sense, is nothing more than a Carnot engine designed to refine raw input energy into highly specific and ordered output energy (highly coherent "in-phase" photons of single-frequency or "monochromatic" light), in order for us to understand how a laser performs this refinement process, we must understand the basic refining components at the heart of the laser. The very complex components used to do

---

[3] See this author's prior work "The Gilligan Principle" for an in-depth discussion on the complex mechanisms found within a living cell (at "EpiphanyBySteveLee.com").

this refinement in a laser are the individual atoms and molecules that form the "Gain Media".

Once we understand how these atoms and molecules interact with the energy we inject into the system, we can then use this knowledge to develop and refine our laser "engine" to increase its efficiency and hence the amount of energy we can successfully transfer into the output product. In other words, armed with the knowledge of how these atomic engines work, we will be able to find ways to help enhance their operation to improve the way in which they refine the energy we inject into them, to enhance their specialized output and thereby improve the operation of the generic laser.

We start by considering one of the most fundamental aspect of all "engines" (and in fact of *everything* we have ever witnessed throughout the entire universe), namely that in the process of transferring or converting energy from the raw input into the refined output, we *always* lose some amount of energy in the process. In other words, no engine is 100% efficient, whether that be a steam engine, laser, animal, plant or ameba. This concept is embodied in what has become known as the *Laws of Thermodynamics.* These laws were gradually developed into their present form through the work of such men as Robert Boyle, James Watt, Nicolas Carnot, Lord Kelvin, and others in the early-to-mid 1800's, forming an essential theoretical foundation that has a fundamental impact on all of our technological expertise today. In fact, these laws are virtually unique in science as a whole, in that while being incredibly simple, they are at the same time *absolute*, with *no known demonstratable exceptions.* Everything we have ever observed, documented or otherwise witnessed follow these fundamental realities of the universe with absolute obedience. Specifically:

1st Law of Thermodynamics: "you can't win". i.e. you can never get more out of any system than you put into it (i.e. the Law of Conservation of Energy put into practice). This law not only tells us we can never get more power out of a machine than we put into it (i.e. no "Perpetual Motion" machines), it also tells us that we can never expect an “input” to become more ordered or structured by *accidental*, *random* events.

2nd Law of Thermodynamics, (A.K.A. the Law of Entropy): “you can't break even”. This law tells us we will *always* get *less* energy out of an engine than we put in, due to inevitable losses in the system. It also tells us that all things in the universe naturally move towards a less energetic and less complex state. This implies that even if we do the work to refine something, it will naturally become degraded over time due to the random forces present in the environment. That being the case, this law tells us that the sum total of the order in the universe is gradually *decreasing*, going towards a more random, disordered state (i.e. towards Entropy).

If for example we send a message from Denver to Dallas via telegraph, wireless, or fiber optic link (for simplicity, a system with no built-in error correction capabilities), we know from the *First Law of Thermodynamics* if we started with a few grammatical errors in the message, by the time it got to its destination, we can never expect random events to somehow fix those errors in transit. Furthermore, by the *Second Law of Thermodynamics / Law of Entropy* we know that random forces present in all environments, will inevitably degrade the structure / order/intelligence imbedded in the message, producing a flawed output by the time it arrived in Dallas. We could carry this example to the extreme, and say that if we then turned around and retransmitted that message to another destination, which then in turn re-transmitted it on to yet other destinations, *ad infinitum*, the random errors that will inevitably creep into the message during each leg of the process would eventually become so severe, that the message will be reduced to little more than a random jumble of letters and symbols after a few thousand or million resends, i.e. a high state of disorder (or “Entropy”).

Though the Law of Entropy tells us that *all* things in the universe left to the random forces around us trend towards disorder, it does *not* tell us that order can never be achieved. *If* someone with a little intelligence is *willing to pay the price* in effort and energy, they can create a more structured output *locally* and thereby offset the natural tendency of randomness, *temporarily*.

If we are clever enough, we can construct a machine that can refine our raw energy to produce a structured output that is more sophisticated and ordered than we had when we started. However, the price we pay will be in energy loss as well as in complexity. In other words, it will be less than perfect in its attempt to convert the raw input energy into the refined output, consuming some portion of the input energy along the way.

As we shall see, lasers are no exception to the Law of Entropy, offering efficiencies that range anywhere from 1% to well over 50% efficient (which however, is still far better than your average '67 impala). The main factor responsible for the inefficiency in the laser is in the way in which we load or "pump" raw energy into the molecules and atoms that form the heart of the laser itself (the "Gain Media"). Since we cannot directly manipulate each and every atom in the gain media as might be possible with a steam engine or clock mechanism, our pumping techniques tend to rely on brute force, reasoning that *statistically*, if we throw enough energy at the specially chosen atoms or molecules, *some percentage* of that energy will be successfully converted into the output form we need.

Armed with a knowledge of how the individual atoms or molecules work in our laser "engine", we can use those insights to improve the efficiencies, such as using carefully chosen gas mixtures and "impurities" which help "channel" and direct a little more of the input energy to our desired wavelength, etc. Though all pumping techniques always turn out to be less than ideal, they work well enough to meet our needs – plus there is something to be said for simplicity, since it often goes hand-in-hand with reliability.

To compensate for the low efficiencies (and thus low yields), we simply increase the proportions: 10% efficiency may not sound too impressive if we start with only a 1 Watt input, but 10% of a 1 Mega Watt input, can't be dismissed so easily. Again, *it all depends on what price we are willing to pay*.

In Babbage's "Analytic Engines", a sequence of planetary gears and linear gear rods (etc.) were used to convert the raw input power and data into a highly intelligent and refined output. In our family car, camshafts, gears, and pistons are used to convert the raw power locked in the chemical bonds of the fuel molecules into the "focused" mechanical and electrical energy the engine produces. In the laser, we exploit the very nature of the atoms and molecules themselves as our refining mechanism to convert the raw input energy into the coherent monochromatic laser light we seek to produce.

When it comes to making a working laser, there are things we can do to help improve its performance (e.g. the method used to inject raw energy into the laser, improve the containment chamber, "tweak" the gasses used in the gain media to improve the efficiency of the transfer of energy to the desired metastable state, etc.). However as it turns out, the hard part of creating the fundamental mechanisms that make lasers possible, nature has already designed for us! All we have to do is select the right atomic/molecular "machine" for the job, and then just let them do what they already do naturally.

By comparison, for the automobile to have been this conveniently available in nature, the engine (or at the very least the individual components such as pistons and camshafts, etc.) would have to have been as common as rocks and sand. For the laser Engineer, such a significant bit of good fortune cannot be overstated, since the design and manufacture of a device by man that could do what nature already has been doing[4] for billions of years would be an infinitely monumental task.

<u>LASERs</u>:

As most sci-fi buffs will tell you, the term "LASER" is an acronym, which stands for "**L**ight **A**mplification by **S**timulated

---

[4] For example, in 1981, NASA [2] found that the thin atmosphere near the surface of Mars routinely undergoes mild lasing due to the high concentration of CO_2. Since nothing is done to enhance this action, these lasing emissions are extremely weak. As we shall see, the CO_2 molecule has a very well-defined metastable state, making it an ideal choice for a laser gain media.

Emission of **R**adiation". As the name implies, the output light from a laser is amplified in a cavity saturated by "energetic" molecules or atoms, which serve as the "Gain Media". As the excited atoms or molecules in the gain media begin to de-excite, they release photons. As the released photons pass by other excited atoms, they trigger or "stimulate" those excited atoms/molecules to release their stored energy in the form of photons, which in turn stimulates still other atoms/molecules around them to release their stored energy in a chain reaction, producing a "coordinated" avalanche emission of radiation in the form of photons. Since the photons being released are *all of the same wavelength* and are all stimulated by identical photons (i.e. monochromatic) passing nearby, all the photons being emitted are *in-phase* with each other, creating what is known as a "*coherent*" light source (where coherence is somewhat analogous to a marching band locking arms and marching lock-step across a playing field).

There are several different types of lasers (e.g. gas lasers, semiconductor lasers, chemical lasers, etc.), but all share the same basic functional elements, including:

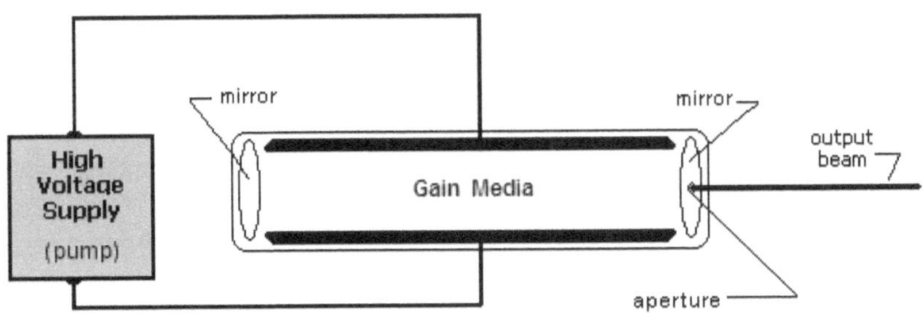

Figure 1.3 Basic elements of a laser "engine".

1) A driving mechanism (or "pump") which injects raw energy into the system (e.g. a high voltage power supply which supplies energy into the gain media in the form of an electron discharge, or flash lamp, etc.),

2) The main laser cavity, which contains the "Gain Media" (e.g. HeNe gas, CO_2 gas, Ruby crystal, etc.) and which

serves to "amplify" the initial few photons released by reflecting them back and forth within the cavity to thus stimulate even more excited atoms in the process,

3) And an output coupling device (e.g. a simple aperture or shutter at one end of the cavity, etc.).

As for the gain media itself, there are two fundamental aspects that are vital to its correct operation, without which lasing and hence lasers would not be possible. These are:

1) The gain media atoms/molecules must "*resonate*[5]" at some particular frequency/wavelength, and thereby perform part of the refinement process of the laser "engine". This resonating aspect of the gain media allows it to absorb specific frequencies/wavelengths of energy from the raw input energy we inject into the system, converting a portion of that absorbed energy into the specific frequency/wavelength we seek for the output emission. However, just being able to resonate at a particular frequency is not enough. All atoms and molecules resonate at some preferred frequency (or frequencies), which explains why one object appears "red" to us, and another "blue", or "yellow", (etc.). However in order for an atom or molecule to make an effective gain media for lasing, it must exhibit one other very essential behavior:

2) The gain media must *store* some portion of the absorbed energy at the desired wavelength for an *extremely* "long" period of time in preparation for the coordinated stimulation of photons. Typically most atoms absorb and then release energy very rapidly, on the order of $\sim 10^{-15}$ seconds, which is far too brief to allow the gain media to setup and store sufficient amounts of energy in preparation for a coordinated, stimulated emission. Atoms or molecules used in the gain media however must have the unique property that they (for reasons we will soon explore) retain that stored energy for an inordinate amount of time, on the order of milliseconds or longer (which is an *eternity* compared to 10^{-15} seconds). This prolonged storage effectively holds the energy of that particular wavelength until it is eventually released in the form of a photon. As the atom or

---

5 See Appendix A: Resonance

molecule emits this photon, it triggers or "*stimulates*" other neighbor atoms/molecules in its path to release their stored energy as photons (of the same wavelength), creating a *stimulated avalanche* of photons. These are then all released in "unison" (*in-phase*) through the cavity aperture, resulting in the strongly *coherent* stream of monochromatic light that is the laser beam itself.

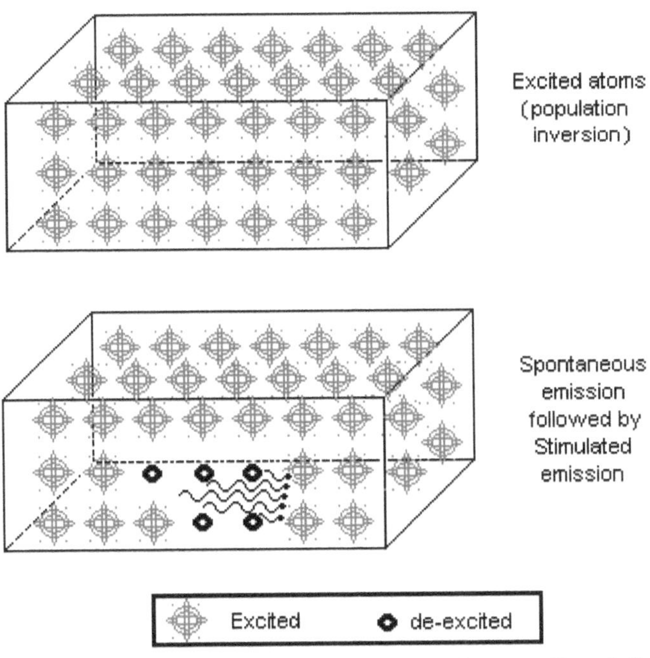

Figure 1.4 Diagram of an inverted gain media, followed by stimulated emissions.

In Quantum Physics parlance, the state where the atoms/molecules hold their stored energy for this extremely long period of time is known as the "*Metastable State*", and can only be understood using models developed by Quantum Physics (to be discussed in Chapters 3 and 4). When we "pump" the gain media by an outside energy source, we place these special atoms/molecules in an *excited* metastable state, collectively referred to as a "*Population Inversion*" (this is considered an "inversion" since all things, including the atoms/molecules in the gain media, naturally seek to be in the more stable de-excited or "ground" state, in keeping with the Law of Entropy). The whole process of "pumping" the gain media and its releasing of its stored energy via Stimulated Emission is referred to as "*Lasing*".

Gain Media	Wavelength	~Pout	cw/pulse
HeNe (gas)	632.8 nm	1 mW	cw
GaAs (semicond.)	0.85^{+} µm	100 mW	cw
CO_2 (gas)	10.6 µm	500^{+}W	cw
CO_2 (gas)	10.6 µm	5 MW	25ns
Nd:YAG (crystal)	1.06 µm	10^{+} W	cw
Nd:YAG (crystal)	1.06 µm	2 kW	50 ps
Nd:Glass (crystal)	1.06 µm	10^{12} W	10 ps
fluorescent dye	300 - 800 nm	10 kW	< 1ps
Ar-F (eximer)	192 nm	10 mW	20 ns
O_2-Cl (chemical)	35 µm	20 kW	cw !

Table 1.1 Common types of laser currently available.

Before we dive into the process of lasing, we first need to develop a simple model that provides a rough idea of how atoms/molecules absorb energy through "resonance". For this initial "first approximation" view of atoms, we will use a classical model based on a simple "ball and spring" design, allowing us to describe both the essential effect of "resonance" demonstrated by all atoms and molecules, as well as the subsequent propagation of laser light from these resonating atoms/molecules.

Chapter 1 Review questions:

1. Explain the basic concepts of a Carnot engine.
2. What aspects of a Carnot engine describe the key functions of a laser?
3. Explain the first Law of Thermodynamics.

4. Explain the second Law of Thermodynamics (Entropy).
5. Explain efficiency of an "engine" in terms of the second Law of Thermodynamics.
6. Other than frictional heating and the desired output, where does all the energy go that is injected into an engine?
7. Explain each term in the acronym "LASER".
8. Explain "*coherent monochromatic*" emissions.
9. Explain the concept of *stimulated* emissions.
10. What refining mechanism(s) do we use in our laser engine?
11. Describe the two fundamental requirements before molecules or atoms can serve as a lasing medium.
12. What is the purpose of the laser "*Gain Media*"?
13. What is a "*Metastable State*"?
14. How critical is the Metastable State to the process of lasing? Why?
15. What is a "*Population Inversion*", why is it an "inversion", and why is it important to lasing?
16. Explain what is meant by the phrase "pumping the gain media".
17. Name two common methods used to pump the gain media.

Chapter 2 The Classical model of an atom.

Atomic resonances and the Lorentz Model:

As mentioned in the previous chapter, a fundamental requirement for the lasing process is that the atoms and/or molecules used in the gain media must *resonate* at the desired wavelength (see Appendix A). As they resonate with this energy, they effectively extract some portion of that energy from the raw power source used to pump the "engine" that is the laser. To fully understand these resonances (and the metastable state in particular), we will need to use the more accurate models developed through Quantum Physics. However, before we jump into the deep end, there is a relatively crude classical description we can use based on a "ball-and-spring" Simple Harmonic Oscillator ("SHO") model of the atom that surprisingly yields a fairly reasonable first approximation to the phenomenon of atomic/molecular resonance (± an order of magnitude, or two). Once we have this SHO "ball and spring" model of the atom under our belt, we can then extend the concept to help understand the resonances of molecules.

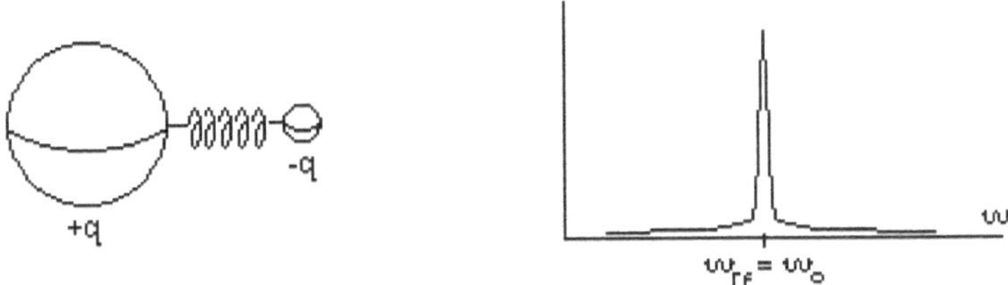

Figure 2.1 The simple "Ball and spring" atomic resonance model.

To begin, we construct a simple model of two charged particles, "tethered" together by a small spring representing the attractive force that exists between two oppositely charged particles. Any time we attach two objects together with some "elastic" or "spring-like" tether, we find that the combined structure tends to vibrate/oscillate at a particular frequency. That particular frequency of oscillation depends on the size (mass) of the two objects, as well as the strength of the restoring force in the "spring-like" tether holding the objects together.

The forces involved in this simple "ball and spring" system are:

$F_{spring} = -k\,x$. The magnitude of the force created by a spring depends on the separation distance "x" and the strength of the spring's restoring force, represented by the force constant "k". This force pushes in the direction *opposite* the separation distance, hence the negative sign.

$F_{electric} = -q\,E$. This is the "Lorentz force", and it describes the force experienced by a charged particle "q" exposed to an external ElectroMagnetic field, where q = negative charge (an electron) and E = Electric field strength. For simplicity we have neglected the magnetic field component, since the Magnetic field is typically several orders of magnitude weaker than the Electric field at normal velocities.

$F_{damping} = -\alpha\,m\,v$. This represents "viscous" damping effects experienced by any object moving through some "viscous" media. This force depends on mass and velocity, with some proportionality constant "α" defining the "resistance" of the viscous dampener. This force also pushes in the direction *opposing* motion, hence the negative sign.

These forces combined must equal the total net force on the ball, which Newton's Second Law tells us equals mass times acceleration (F = m a):

$$F_{total} = -\alpha m v - k x - q E = m a$$

We note that velocity is the first derivative of "x" (with respect to time), i.e. the amount of *change* in distance per second. Similarly acceleration is the first derivative of velocity (i.e. the amount of *change* in velocity per second), making it the *second* derivative of "x". Therefore we can rewrite this equation as:

$$F_{total} = m * d^2x/dt^2 = -\alpha m * dx/dt - k x - q E$$

We note from High School physics that the frequency of oscillation of a spring system is proportional to the strength of the spring constant ("k"), and inversely proportional to the mass the spring has to move, i.e. $\omega_o = SQRT(k/m)$ (where ω is the angular frequency, and is related to the linear frequency, $F=2\pi\omega$). Using this, we replace "k" with $m\omega^2_o$, and rearrange terms:

$$m [d^2x/dt^2 + \alpha * dx/dt + x \omega^2_o] = -q E$$

Since we know this system is an oscillator, we expect the solution for "x" (the masses' position) to look periodic. Therefore we try a solution of the form[6]: $x = A \exp(-i\omega t)$. We now take the derivatives of this solution and plug them back into our equation (noting that $[-i]^2 = +1$):

$$dx/dt = -i \omega A \exp(-i \omega t)$$

$$d^2x/dt^2 = (-i)^2 \omega^2 A \exp(-i \omega t)$$

Therefore:

$$m [\omega^2_o - \omega^2 - i \alpha \omega] A \exp(-i \omega t) = -q E$$

We note that "$A \exp(-i\omega t)$" is our original expression for "x" and therefore we can rewrite the above equation as:

$$m [\omega^2_o - \omega^2 - i \alpha \omega] x = -q E$$

We now introduce the dipole moment "p":

$$p = q x$$

---

6 We note that this is a sinusoidal function, as demonstrated by "Euler's relation" (equation D.1 in Appendix D).

which is simply the charge multiplied by the distance separating the two charged particles (we will find in Chapter 6 that in order for a charge distribution such as a dipole to radiate, those charges must move, or in this case, oscillate). Rearranging terms to specifically introduce the dipole moment we get:

$$q\,x = -\,q^2\,E\,/\,(m\,[\,\omega^2_o - \omega^2 - i\,\alpha\,\omega\,]\,) \qquad \text{Eqn 2.1}$$

Looking at equation 2.1 for a moment, we note that when the last two terms in the denominator ($-\,\omega^2 - i\,\alpha\,\omega$) combine to equal the square of the "spring" resonance frequency ($\omega_o{}^2$), we get a singularity (i.e. 1/0), which implies the dipole moment becomes extremely large anytime the external field frequency (ω) approaches the resonant frequency of our atom or molecule (ω_o), creating a pronounced spectral peak at that frequency (shown on the right half of Figure 2.1).

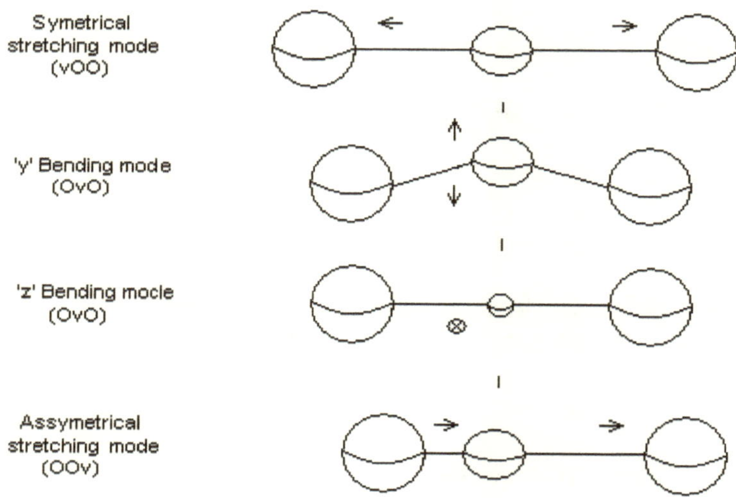

Figure 2.2 The various vibrational modes of the CO_2 molecule.

For more complex systems with more than two charged particles, we expect that many such resonances will exist. However, in such complex systems, the combined fields of all the near neighbor atoms complicate the system significantly, making it much more difficult to model with any degree of accuracy. This Simple Harmonic Oscillator (SHO) model however, does allow us to demonstrate the general resonant behavior exhibited by atoms when activated by an external energy source (e.g. an external electromagnetic field, collision with an electron "arc", etc.).

By extension, we can develop a similar SHO model to describe simple molecules, since they too are composed of charged atoms held together with a similar "elastic" / "spring-like" force developed by the electrostatic attraction between the charges involved. For example, a carbon dioxide CO_2 molecule[7] is composed of three atoms held together by two electrostatic "spring-like" forces between them. When struck by some external energetic object (e.g. the energetic electrons in an electron discharge "arc"), the three atoms begin to vibrate. There are several different vibrational modes in such molecules, including a bending vibration as well as a longitudinal axis vibration (see Figure 2.2).

A principle known as the "Kramers-Kroenig relationship" tells us that there is a direct correlation between the atomic/molecular *resonances*, and their subsequent *absorption* of the energy impinging on them from an external Electro-Magnetic field. This inherent resonance and absorption at the oscillating frequency of the "ball and spring" atom/molecule effectively defines how external fields propagate through material composed of these atoms/molecules. We account for this material response to the Electric field using what is known as the material's "permittivity" (ε), while its response to the Magnetic field is similarly quantified using its "permeability" (μ).

After these atoms/molecules have absorbed some of the energy of the external field, they at some point then release that

---

[7] We note that CO_2 and Nitrogen are two of the more popular choices for a basic "home brew" laser project. Though it is easy to get these gases to lase, it requires a high voltage to pump the gain media. CAUTION: Such high voltage can cause permanent nerve damage, and can be fatal.

energy in *random* directions (i.e. typically not in the same direction the original field was traveling). This selective absorption and release of certain frequencies (or "colors" of light if in the visible spectrum) creates an effect on that field generally described as "Chromatic Dispersion", in which different frequencies (or "colors") of light are affected *differently* by the media in which they propagate, changing their amplitude, phase and/or timing relative to each other in the process.

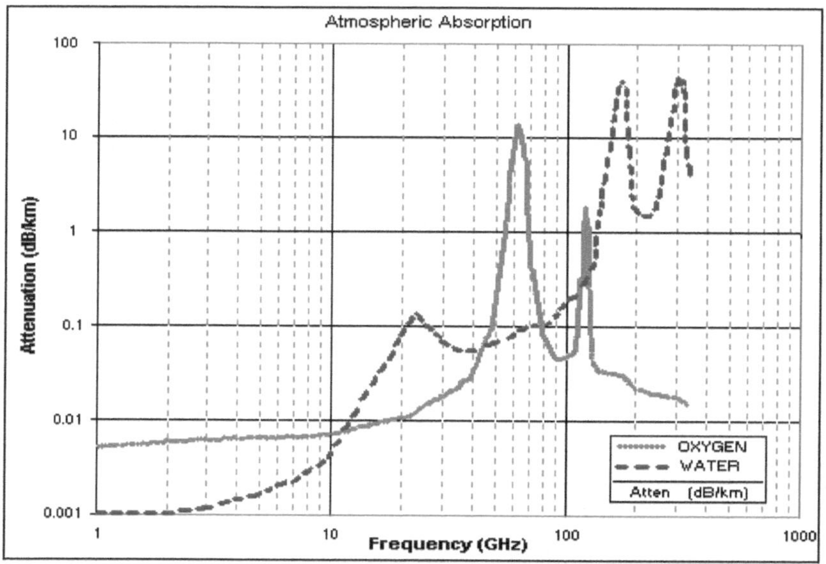

Figure 2.3 Sample spectral absorption plot, atmospheric H_2O and O_2

A common example of this effect is found in the sky over our heads. Earth's atmosphere is composed of ~78% Nitrogen, ~17% Oxygen and ~3% water (plus other trace gases), which gases tend to resonate strongly with light in the blue region of the visible spectrum. When white light (i.e. light containing "all colors of the rainbow") from the sun hits our atmosphere, these gases absorb the blue light, and then shortly thereafter re-radiate it in random directions, giving our sky its characteristic blue tint. This also explains why the sky at sunrise and sunset displays more red/orange light, as most blue light is scattered out and thus attenuated after passing through "twice" as much atmosphere when the sun is at such low angles.

Similarly these gases also have several pronounced absorption peaks in the microwave range related to Oxygen and water vapor (see Figure 2.3), that can also be understood conceptually using this "ball and spring" resonance model.

The bottom line to all of this as far as lasers go, is that atoms/molecules in our gain media (as well as in the path of any laser beam we generate), will first resonate with and thus absorb some portion of the energy we throw at them, and later release that energy. If we hit them with a "broadband" energy source (i.e. energy with a wide range of frequencies such as in an electron arc, or a flash strobe, etc.), the atoms/molecules in the gain media will select out only specific frequencies that correspond to their resonant modes (i.e. frequencies that equal the "ball and spring" oscillation frequencies of the atoms/molecules involved). They will then at some later point release that energy at the specific frequencies directly related to these resonant modes.

This selective absorption mechanism allows us to hit our gain media with a burst of raw energy (that is typically very "broad band" in nature), from which the gain media will then extract energy at its resonant frequency(s). In the case of the CO_2 laser for example, we hit the CO_2 gain media with an electron discharge. The CO_2 molecules are thus excited through direct collisions with the electrons in the discharge, absorbing energy in their resonate modes in the process (see Figure 2.2). CO_2 gas makes an ideal choice for a gain media, in that it not only has a well-pronounced metastable state (at 10.4 μm), but it is also relatively easy to obtain, and thus makes a "fairly" inexpensive laser. The lifetime of the metastable state itself is on the order of 1 μS, offering plenty of time to establish a population inversion and subsequent stimulated de-excitation of coherent light. We will revisit each of these key aspects of lasing in greater detail as we progress through the following chapters.

Chapter 2 Review questions:

1. Describe the "ball and spring" Simple Harmonic Oscillator model of the atom.
2. What is Permittivity? What is Permeability?
3. What is Chromatic Dispersion, and what causes it?
4. Explain atomic resonances in terms of the simple Lorentz "ball and spring" model. Why is it important to lasing?
5. What causes an atom or molecule to behave like a "Simple Harmonic Oscillator"?
6. Though resonance is key to lasing, is it enough to create a working laser? Explain.
7. Describe the vibrational resonance modes of the CO_2 molecule.
8. How do we exploit this selective absorption of energy in setting up the lasing process (i.e. pumping the gain media)?
9. Given a "N=N" molecular "bond-spring" with a bond strength ("k") of 2297 N/m, and a mass of 1.162×10^{-26} kg, find the resonant frequency of vibration (ω_o). (7.1×10^{13} Hz)

Chapter 3 From Classical to Modern Physics

At the turn of the 20th century, there was something of a prevailing attitude of over-confidence amongst most academics of the day who taught students that science "knew everything there was to know", and had in fact solved all the mysteries of the universe. The only things yet to be done, they told students, was to refine the accuracy of the measurements already made and the universal constants embedded in the equations codified by the likes of Maxwell and Kelvin up to that point.

Oddly enough, at the same time there existed a number of annoying little "anomalies" that kept showing up in various laboratories scattered about, that dared to undermine mankind's claim to omniscience. This misfit data came in the form of "anomalous" spectra emitted by various gas lamps that despite the insistence of academics, refused to produce the continuous "rainbow" everyone had come to expect, exhibiting instead an assortment of rather "odd" disjointed spectral lines. This miscreant data no sooner came to be established as fact, when yet another rather odd (not to mention lethal) effect was discovered by the Curries which radiated off certain heavy elements such as Radon and Uranium, and which imprinted images on "well-protected" photographic film.

And if these anomalies weren't enough, there was that annoyingly defiant problem of modeling the spectral emissions of "Black Bodies", which stubbornly *refused* to take on the nice symmetrical shape of the standard Gaussian distribution to which virtually all other *reasonable* phenomenon comply. Yet through it all, the unflinching arrogance of man continued to insist that nature was in errs, and that science was supreme, pesky little "anomalies" notwithstanding.

In an attempt to work around the Black body conundrum, a guy by the name of Max Planck decided to try what he thought at the time was a trivial little mathematical "*trick*". This "trick" was to treat a Black Body object (or energy radiating off a white hot object, etc.) not as *one* single source of energy, but as a collection of an *infinite* number of little discrete microscopic oscillators, each operating at slightly different wavelengths. His intent was to use this little "gimmick" to work through to a final solution, and then go back when he was done and "simply" replace all the discrete microscopic oscillators with one single source.

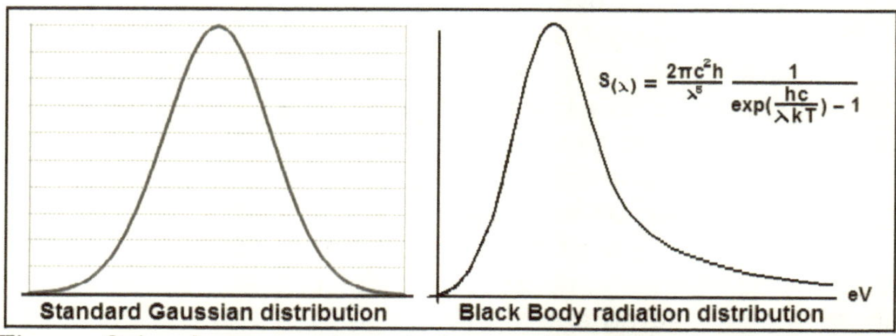

Figure 3.1 A comparison between the "Standard" Gaussian distribution of energy, and that of Black Body Radiation (including Planck's Black Body solution).

Using this approach Planck got his solution, and it did indeed provide a beautiful curve that matched the Black Body's spectral profile, however when he went back and attempted to remove the discrete aspects of his model, he found doing so was virtually impossible. With hindsight, the reason is clear: single objects, such as a Black Body, or a white hot chunk of iron, are not really single objects, but are in fact composed of a near infinite number of individual atoms. Each of these individual atoms are in fact discrete microscopic sources whose combined radiating energy form the total spectra that our eyes and measuring instruments *average* into a single composite or "continuum".

In using his little "trick" to reach the solution to the Black Body problem, Max Planck opened the door to the concept of descritization. To us now this concept seems perfectly reasonable in hindsight, but at the time it was viewed almost as heresy by those trained in classical physics. Soon this "trivial" little trickle of change rapidly expanded into what became a deluge, the likes of which had not been seen before nor since in any field of science. In its wake, it not only brutally exposed the delusion of man's omniscience for what it was, it also in the same stroke up-ended vast tracts of classical "hallowed ground" which those who understood its implications, could scarcely bring themselves to consider. In this single stroke, it revealed the fact that many of even the "simple" classical problems, were in fact infinitely more complex than people at the time admitted, with many of these problems being well beyond the reach of the "deterministic" mindset of classical science (e.g. the four billiard ball collision problem), ripping gaping holes in the veil of man's self-aggrandizing delusions in the process.

Einstein further opened the door to this "Pandora's box" of the Quantum world in his work on the photoelectric effect (i.e. the ability of light to induce small currents in some materials). In his treatment of this problem, Einstein similarly applied the quanta approach to light, describing light energy not as a wave, but as a barrage of individual microscopic "particles" of light he termed "Photons". It was this ground-breaking thesis (for which he received his one and only Nobel prize in 1905), coupled with Planck's descritization of matter, the proverbial mustard seed of revolution in science developed, and from which the sequoia of Quantum machinery came crashing through.

Oddly enough, as Einstein later wrestled with the bizarre and paradoxical oddities that became almost "part and parcel" of Quantum Mechanics, he became something of the heretic to this new discipline, rebelling against the rip-tide of Quantum euphoria which was exploding onto the scene. Most of his resistance to the Quantum approach, was due to his heavy indoctrination in the notions of classical determinism, and its ingrained over-confidence in its ability to solve all problems *exactly* and with *absolute* finality.

Einstein's oft quoted objection that "God does not play dice with the universe" underscored his belief that, given enough time, any *real* scientist would be able to tell you with absolute certainty exactly what every molecule in a gas will do, and would not have to "settle" for the Quantum Mechanical approach of estimating the statistical *probability* of what the gas "might" do.

It is one of the great ironies of history that Einstein became simultaneously both the greatest impetus, and one of the greatest critics to this new branch of physics based on statistical probability and bizarre paradoxes, almost acting as a human analogue to the nature of the Quantum Physics itself. Even in his efforts to expose Quantum techniques for the heresy he perceived them to be, he both helped refine this discipline through its infancy, as well as helped define its ambiguity by offering several yet unresolved paradoxes that have become *the* perennial topics of debate routinely visited by physicists and sci-fi enthusiasts alike, almost in obligatory prose.

Einstein's theory of light and atoms

Absorption:

When an electron in an atom is exposed to an external source of energy (e.g. a flash of white light, or the kinetic energy of an excited electron by collision, etc.), the amount of energy that an electron will absorb (and hence store in the atom it is attached to) is equal to the difference between the "orbit" (or energy level) it was in, and the "orbit" it is elevated to:

$$\Delta E = E_i - E_j$$

where i and j represent the ith and j^{th} energy level of the atom.

In other words, when an atom absorbs energy from the external raw energy source, it absorbs *exactly* the amount that is needed to elevate it to its new "orbit" around the atom. This absorbed energy is equal to the difference in energy between these two energy levels in the atom. Once an electron is excited into a higher (less stable) orbit, its elevated state effectively represents energy stored in that atom. At some point the atom will release this energy as the electron relents to this attractive force and falls back to the lower (more stable) orbit.

Consider the following argument by analogy: if we pump 200 gallons of water up a hill into a tank against the force of gravity, will that water have potential energy stored in it? Yes.

Where did it come from? From us; we put that energy in it – or more correctly, in the gravitational "system" of masses composed of the water and the Earth – when we fought the pull of gravity and elevated the water up the hill (PE = Force*Distance = mg*H, where m=mass, g=the acceleration of gravity=9.8 m/s^2 and H=height). We can easily prove there is energy stored in this "system" by letting the water run downhill and over a watermill to generate electricity, or grind flour, or cut wood (etc.). All of these tasks represent work being done, and no work can be done without some energy to drive the process – energy that therefore had to have been stored in the water atop the hill.

Example: Water and the force of gravity

How much energy is stored in 1000 kg of water lifted vertically 100 meters against the pull of gravity? How much power is released if all of that water were released and converted into work in 60 seconds (assuming an ideal zero-loss process for simplicity)?

$$\begin{aligned} PE &= F*Distance = ma * H \\ &= (1000\ kg)\ (9.8\ m/s^2) * (100\ meters) \\ &= 980{,}000\ Joules \end{aligned}$$

Power = work done per second (Watts = Joules/Second), therefore:

$$Power = 980{,}000\ J / 60\ seconds = 16.3\ kW$$

<u>Metastable states</u>:

As mentioned previously, most all atoms and molecules tend to de-excite very rapidly (on the order of ~10^{-15} seconds). However it has been known since ancient times that some minerals can store energy in an excited state for a much longer period. This prolonged energy storage state became known as a "Metastable state", and can have a "lifetime" that can range anywhere from milliseconds (referred to as "Florescence"), to minutes (referred to as "Phosphorescence") with some compounds. The classical Lorentz "ball and spring" model cannot explain this behavior, since according to classical physics all such "loaded" spring systems should immediately release their stored energy as soon as the "loading" force is gone. However experience with phosphorescent and florescent material clearly shows this model to be wrong in these cases, leaving classical physics with no valid way to explain this "anomalous" behavior.

To explain this prolonged energy storage mechanism seen in some atoms and molecules, we must draw upon models developed using the "discrete" picture described by Quantum Physics (discussed below). Suffice it to say that without this discrete Quantum model, an explanation for these Metastable states, and hence an understanding of the laser gain media itself, is not possible. And without that understanding, neither can we understand the existence of the "population inversion" in the gain media, leaving us with no way to develop or refine a functional laser.

<u>Spontaneous Emission</u>:

Once we have successfully elevated atoms into a valid metastable state, we find a few of these atoms eventually de-excite on their own. For an atom to de-excite, it must discard exactly that amount of energy that represents the difference between the state that it is in, and the energy level that it seeks to be in. To discard this energy, the excited atom ejects a photon that is equal to this difference in energy:

$$E_{photon} = E_i - E_j$$

In other words, as the electron in the excited level falls into the lower energy level (nearer the positively charged nucleus that attracts it), it discards the excess energy in the form of an ejected photon (this is somewhat analogous to what an orbiting space shuttle must do in order to return to the Earth, i.e. it must reduce its kinetic energy by blasting "retro rockets" in the forward direction. As it slows, it loses kinetic energy and descends in altitude, giving up potential energy "stored" in its higher altitude).

The energy of the ejected photon is directly related to its frequency:

$$E_{photon} = hf$$

where h = Planck's constant = 6.626×10^{-34} J-sec

Stimulated Emissions:

If the excited atom happens to be in the company of other excited atoms (i.e. in the midst of a population inversion), when it ejects this photon, that photon inevitably passes one of these excited neighbor atoms and in the process stimulates it to de-excite. Since both atoms are identical in composition (i.e. the same number of protons), with identical energy levels (and hence shed identical amounts of energy when they de-excite), the energies of the photons they both eject are equal. Also, since the energy of the photon is directly related to the frequency/wavelength of the photon via the equation above, both photons have exactly the same frequency/wavelength, and therefore together they form the beginnings of a stream of "monochromatic" light.

What's more, since the second photon was stimulated by the near proximity of the first, the second photon leaves its atom *in-phase* with the first (triggering) photon. Consequently when they travel together, they are not only both of the same wavelength, but are also *in phase* (i.e. "coherent", analogous to an army of photons marching "lock-step" downstream). It is this monochromatic coherence that makes laser light different from all other light sources, in that all other types of light sources emit photons that have no phase relation to each other, and as such do not "reinforce" one another).

As these two coherent photons then travel on through the "inverted" gain medium and pass by other excited atoms, they in turn stimulate yet other photon emissions, each of which is released with the same wavelength and phase as the first. If sufficient numbers of atoms in close proximity start this process in an excited state, the stimulated avalanche can generate perhaps ~10^{20} or more coherent photons in less than microseconds.

From the mountain of available data (e.g. spectral plots), it is clear that the amount of energy difference between two levels in an atom is *discrete*. In other words, in the absorption and emission process, the energy of the photon absorbed or emitted must be *exactly* equal to the difference between these two *well-defined* energy levels – *no more*, and *no less*. This tells us that energy levels or "orbits" in-between these well-defined levels are not allowed by the atom, and are therefore considered by Quantum Physics to be "forbidden".

We know that the charge on every electron is always the same: 1.602 x 10^{-19} coulombs, no more, and no less. What's more, the energy stored in the electron orbits depends directly on the electron charge and the attractive force between the electron and proton holding its mass in orbit. Since the attractive force is directly dependent on the charge of each particle, and since the charges on an electron and on a proton are specific, *distinct* values, it is logical to assume that the energy levels formed in the atomic system therefore define very distinct and *discrete* values.

In other words, since electrons *always* have *exactly* the same mass and charge, and protons likewise have exactly the same value of charge (though with the opposite polarity), the radius of the orbit of an electron (held in place by the Coulombic attraction between these two charges) always takes on specific and *discrete* values for each type of atom.

The key difference between one type of atom (e.g. Hydrogen) and another (e.g. Helium, or Carbon, etc.) is in the number of protons at the center of each atom. Though other elements do have more protons, and thus generate more Coulombic attraction on the electron towards their nucleus, the number of additional proton charges is an *integer*, implying the total value of their accumulated charge is *quantitized* into discrete "steps" of charge (e.g. *two* protons together in the nucleus produces a net charge of 2x 1.602 x 10^{-19} coulombs;

three protons in the nucleus would have 3x 1.602 x 10^{-19} coulombs, etc.). Therefore the attractive force between the nucleus and the electrons orbiting it is likewise quantitized. As we shall see, this leads to energy levels or "orbits" with very discrete radii.

The fact that all atomic systems form very *discrete* energy levels, implies *only* those photons that correspond to one of those *discrete* energy states can resonate and thus be absorbed or emitted by an atom (as we shall see later, a similar argument can be made for molecules, since they are likewise a composite of discrete masses and charges).

For these reasons, *all* transitions that do not correspond to one of these specific and discrete energy levels are "*forbidden*". These conclusions are very clearly supported by spectral plots, which never show normal/stable atoms taking on energy corresponding to anything but the allowed energy levels.

In other words, a given type of gas (e.g. Hydrogen, Helium, Methane, etc.) *always* have the same energy levels/orbits, and thus always exhibit exactly the same spectral "signature". Not only do these spectral "signatures" always have their spectral lines at the same wavelengths, they never show any energy (spectral lines) at any "forbidden" energy level. These facts lead us to conclude that all atomic systems (and consequently all molecules) are *quantized*, and this quantized nature can only be accurately understood using Quantum models which specifically account for this discrete aspect of matter. In fact, it was the remarkably good correlation between the predicted location of these spectral lines using Quantum models, and the actual spectral data that provided some of the strongest validation for the branch of science that became known as Quantum Physics .

<u>The Bohr Model</u>:

Following Einstein's quantitization of energy related to atoms, the question remained as to what these energy levels looked like or represented in terms of the actual atomic structure. In 1913 Niels Bohr attempted to answer that question by proposing a model of the Hydrogen atom[8] describing it as something of a microscopic planetary system, with the nucleus at the center and electrons orbiting around it at *fixed radii* from the nucleus, analogous to the way planets orbit the sun.

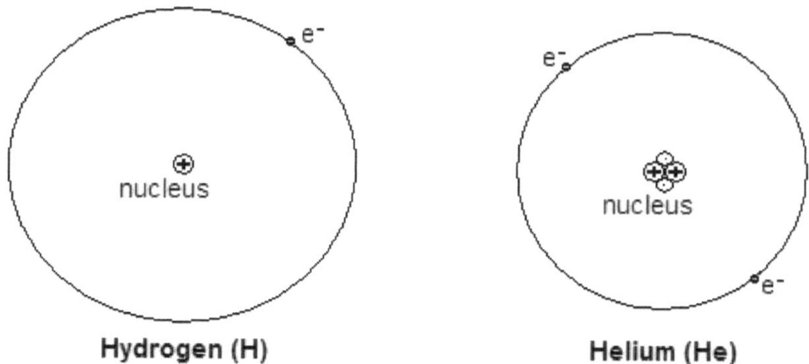

Figure 3.2 Bohr atom approximation for Hydrogen and Helium (not to scale)

In the case of a solar system, planets are held in orbit by the force of gravity between their mass and that of the sun. In the atom, the orbiting electrons are similarly held in orbit by the attractive electrostatic force between the electron and the positively charged nucleus. Though this model is something of a crude "stick figure" first approximation, it was the first to reasonably explain the spectral lines produced by glowing gas lamps seen by many 19th century scientists, validating the basic

---

8 For reference: The Hydrogen atom in this model is estimated to be about 10^{-10}m in diameter, the proton ~10^{-15} m in diameter, and the electron ~10^{-18} m in diameter, hence the ratio between the total area covered by the electron's outer "orbit" (the atom's diameter) verses the diameter of the proton is ~10^5! Since the electron's size is so miniscule compared to the total diameter of the atom, we conclude that there is a great deal of empty space around the atom that is "unoccupied" by the electron (though this description of the atom is only a simplified approximation, this general conclusion remains valid; see discussion in Chapter 4).

elements of the model. Bohr created this model by extending classical concepts, while applying Planck's constant "h". This effectively imposed a *discrete* quantization on the atoms, giving rise to the *discrete* lines in the gas spectra.

To define the atomic orbits in his model, Bohr started with the Lorentz's force (see Chapter 2) between the electron and the proton, along with Coulomb's law to define the static Electric field generated by the proton at the nucleus of the Hydrogen atom:

$$F = q E = e e /(4 \pi \varepsilon r^2)$$

(where "e" = 1.609×10^{-19} coulombs of charge). As the electron orbits the nucleus, this Coulombic force holds the electron in, and in classical terms is known as the "Centripetal force". Classically, a centripetal acceleration is defined as:

$$a = v^2 / r$$

If we combine this acceleration with Newton's second law: (Force = mass * acceleration, F = m a)[9], this becomes:

$$F = m v^2 / r = e^2 / (4 \pi \varepsilon r^2)$$

Canceling "r's" we get:

$$m v^2 = e^2 / (4 \pi \varepsilon r)$$

Since Kinetic Energy is: $\frac{1}{2} m v^2$ we can multiply through by ½ to show the Kinetic Energy in the system:

$$\tfrac{1}{2} m v^2 = e^2 / (8 \pi \varepsilon r)$$

Potential energy "U" in any such system = Force * distance separating the two charges, or:

---

9 To be technically correct, "m" here is the "reduced" mass of the electron-nucleus system, i.e. the mass as if it is all located at the center of gravity: $m = m_e m_N / (m_e + m_N)$

$$U = (+e)(-e)\,r \,/\, (4\,\pi\,\varepsilon\, r^2) = -\,e^2 / (4\,\pi\,\varepsilon\, r)$$

Therefore, the total energy in the system, E_{total} = Kinetic + Potential:

$$E_{total} = e^2 / (8\,\pi\,\varepsilon\, r) - e^2 / (4\,\pi\,\varepsilon\, r)$$

$$= -\,e^2 / (8\,\pi\,\varepsilon\, r)$$

Classically, the angular momentum of a spinning object is: $L = m\,v\,r$, However, if we now follow Bohr's [arbitrary] assumption that the angular momentum is quantitized (as shown below) and combine the two forms:

$$L = n\,h/(2\,\pi) = m\,v\,r$$

Manipulating our original Force relation, we can construct an "m v r" term by first multiplying both sides by "m":

$$m^2\,v^2 = me^2 / (4\,\pi\,\varepsilon\, r)$$

Taking the square root and multiplying through by "r":

$$m\,v\,r = \mathrm{SQRT}\,[\, me^2\, r / (4\,\pi\,\varepsilon\,)\,]$$

Equating this now with our angular momentum, we get:

$$L = n\,h/(2\,\pi) = \mathrm{SQRT}\,[\, me^2\, r / (4\,\pi\,\varepsilon\,)\,]$$

Finally, we solve for "r" to find the allowed orbits:

$$[\,n\,h/(2\,\pi)\,]^2 = me^2\, r / (4\,\pi\,\varepsilon\,)$$

Therefore:

$$r = n^2\,h^2\,\varepsilon / (\pi\, m\, e^2)$$

We can condense this down by substituting in the known constants:

$$r = n^2 R_{bohr}$$

where $R_{bohr} = h^2 \varepsilon / (\pi m e^2) = 0.0529$ nm

From this we can calculate the Bohr radius (approximation) for the first few orbitals:

$$r_1 = R_{Bohr} = 0.0529 \text{ nm}$$

$$r_2 = 4 R_{Bohr} = 0.2116 \text{ nm}$$

$$r_3 = 9 R_{Bohr} = 0.4761 \text{ nm}$$

Plugging our radius equation into our "E_{total}" equation above, we can solve for the approximate energy values of each orbital:

$$E_{total} = - e^2 (\pi m e^2) / (8 \pi \varepsilon \; h^2 n^2 \varepsilon)$$

$$= - e^4 m / (8 \varepsilon^2 h^2 n^2)$$

Once again, since these constants are known, this expression can be condensed into the following form[10]:

$$E_{total} = - 21.675 \times 10^{-19} \; C^2 \; V^2 / J$$

$$= -13.53 \text{ electron-Volts} / n^2$$

where one "electron-volt" ("eV")[11] is the energy required to move one electron charge through a one volt potential: 1 eV = 1.602×10^{-19} J.

---

10 Note: 1 volt = 1 Joule / Coulomb

11 Any charge subjected to an electric field, experiences a "Lorentz" force = charge * voltage, much the same way something lifted off the ground experiences a gravitational force. The potential energy

Exercise:

Use the above equation to find the photon energy produced when an electron goes from the 3rd and 4th energy level, to the 2nd level in the Hydrogen atom.

$E_{3,2}$ = – 13.53 electron-Volts ($1/3^2$ – $1/2^2$) = 1.88 eV

$E_{4,2}$ = – 13.53 electron-Volts ($1/4^2$ – $1/2^2$) = 2.54 eV

Exercise:

What are the wavelengths of these released photons? Since E = h f, and λ = c/f, the wavelength λ = c h / E:

$\lambda_{3,2}$ = (300×10^6 m/s) (6.626×10^{-34} J-sec) / 1.9 eV
= 658 nm (Red)

$\lambda_{4,2}$ = 487 nm (Blue-Green)

Again we note that the correspondence between the wavelengths predicted by this model and the actual spectral data measured in labs everywhere, proving the validity of both this model, and the discrete quantized nature used to derive it.

---

in both cases is equal to the work done in lifting the object against the force, i.e. PE = U = Work = F*distance. Therefore the potential energy of an electron subject to a voltage ("V") is: U = q V.

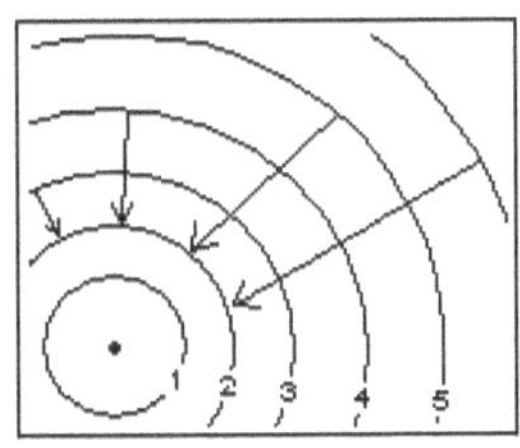

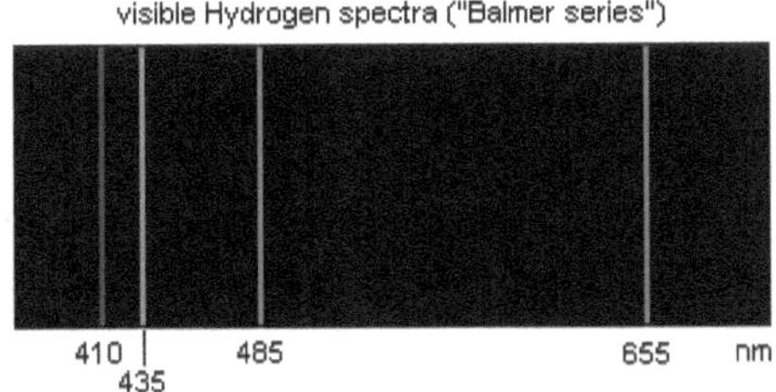

Figure 3.3 Visible Hydrogen spectra ("Balmer series").

<u>Young's Double Slit experiment</u>:

As scientists pondered the spectral data and its implications about the nature of electrons "orbiting" around the atom, someone got the bright idea to repeat an old experiment used roughly a hundred years earlier to examine the nature of light, to help explore the nature of the atom. The results of this experiment were so astounding, that it revolutionized the concept of matter itself.

The original experiment was conducted by a man named Thomas Young, who in the early 1800's, was trying to use it to answer the question as to whether or not light was a *wave*, or a *particle*. Thomas Young was a physician who, in his spare time, managed to shed a great deal of "*light*" not only on the nature of light, but on a number of other puzzling questions of his day, including Egyptian hieroglyphs, and Thermodynamics (obviously, a very busy guy). In 1801, Young used a quasi-monochromatic light source (e.g. light from a Mercury vapor lamp) which he focused on a mask with two slits. Opposite the slits he placed a screen, causing light from the source to pass through the two slits and fall on the screen.

If light were a particle (as Newton had argued), it should fall on the screen in *two* well-defined rectangular patterns, mimicking the outline of the two slits. If light were a wave, it would flow through and around the slits according to Huygen's wave diffraction model (depicted in Figure 6.9), to form a diffraction pattern all across the face of the screen. This would be very analogous to how an ocean wave flows through and around various obstacles in a bay, recombining somewhere past the obstacles and forming large peaks and troughs when the various wave pieces alternately reinforced and canceled each other out.

When Young performed his experiment, he found light sent through the two slits did indeed create a complex diffraction pattern on the screen *in-between* and *on either side* of where the slits would cast their outline, proving light most definitely traveled like a wave. His results were so impressive that this evidence became the "final nail" in the coffin to Newton's proposed "corpuscular theory of light" (and good thing too, what with a name like that).

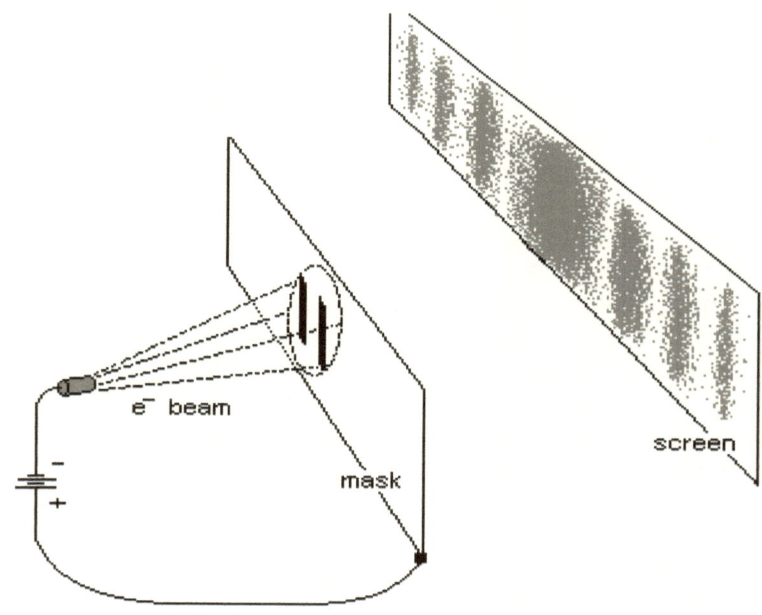

Figure 3.7 Young's double slit experiment using electrons.

In hindsight, we now know from both this experiment and Einstein's work on the Photoelectric Effect that light, like matter, has a *dual* nature to it, which our limited grasp of reality based on limited senses, cannot quite equate to anything in our common, macroscopic world.

What's more, if we repeat this experiment using electrons (or protons, or atoms, etc.), we find that instead of the well-defined outline of the slits that a barrage of "b-b" like particles would cast on the screen, the electrons create a well-defined diffraction pattern. Such a diffraction pattern very clearly indicates that electrons also possess wave-like characteristics much the same way light does. This evidence of the wave-like nature of atomic particles, has profound implications on what electrons, atoms, molecules, etc. are, requiring scientists to adopt analysis techniques previously used to study classical waves, including tools developed by men like Fourier in the 1800's (see discussion below).

<u>Basic principles of the Quantum model:</u>

So far in our discussion, we found that when we have attempted to use classical models alone to explain how the atom works (e.g. Newton's laws of physics, the Lorentz "ball-and-spring" model, etc.), the fit has been "ball park" at best. However, as people such as Planck, Einstein and Bohr began to realize that we need to treat these sub-microscopic objects not as classical thinking might suggest, but as the *discrete* Quantum objects they are, complete with some yet-undefined "wave-like" properties to them, they began to realize that our macroscopic perspective of the universe was incomplete. In a sense, the classical perspective was a "low resolution" view of the universe in which all of the very fine atomic detail had been "glossed over", effectively "averaged out" by the coarse treatment given them up to that point.

In order for us to fully understand the universe built on such a quantatized and "wave-like" reality, we needed to refine our perspective and shift our thinking to accommodate the discrete subtleties of the atomic world. To that end, we now introduce several of the more fundamental Quantum paradigms as "Postulates" used to shape the Quantum Mechanical approach to understanding the universe:

Postulate # 1) Contamination by observation:

Objects in the Quantum world are so tiny, that even the most minute amount of light used to view them profoundly disturbs their state. Consequently, if we attempt to "see" what state an atom or one of its electrons has taken by shining light on the atom, we completely alter whatever state it was in, making any such direct attempt to study the atom's behavior impossible. This is analogous to attempting to attach a cheap oscilloscope probe to a radio frequency oscillator (for those familiar with electronics), in that the act of attempting to observe/measure the oscillator "contaminates" the system by changing the characteristics of the circuit. This contamination by observation typically either changes the output frequency of the oscillator, or kills it altogether.

In the Quantum application of this principle the problem is even more acute, since in order to observe any object accurately, we need to use light or instruments smaller than the finest detail we wish to measure (by analogy, you can't resolve the ridges and valleys in your fingerprint using a probe the size of a bowling ball). Since the shortest wavelength of visible light (near ultraviolet, $\sim 10^{-7}$ m) is *much* larger than an electron (10^{-18} m), normal light is clearly far too large to be used to resolve any of the fine detail of an atom.

Unable to apply the "standard" observational techniques normally used in labs, to study electrons around an atom, we are forced to study the atom indirectly using a *mathematical* model. Since the atomic system has a clear spherical geometry to it (i.e. a spherical cloud of negatively charged electrons orbiting a positively charged nucleus), we find we need a model that has a decidedly *spherical* geometry to it:

Example: Attractive force in the Hydrogen atom.

How great is the force of attraction between the Hydrogen nucleus (a single proton) and an electron in the first Bohr orbit?

Using the Lorentz force ($F = q E$) and Coulomb's Law for the Electric field:

$$F = q E = e \; e / (4 \pi \varepsilon \; r^2)$$

$$F = (1.6 \times 10^{-19} \text{ C})^2 \; / \; [\, 4 \pi \varepsilon \; (.05 \times 10^{-9})^2 \,]$$

$$= 92.6 \times 10^{-9} \text{ N}$$

Example: Attractive force in Iron atom.

Assuming an extremely simplified case where all but one electron has been removed from an Iron atom (Atomic number "N" = 26) at the first Bohr radius, what amount of Coulombic force holds it in?

$$F = (N e) * E = Ne \; e / (4 \pi \varepsilon \; r^2)$$

$$F = 26 \times (1.6 \times 10^{-19} \text{ C})^2 / [\, 4 \pi \varepsilon \; (.05 \times 10^{-9})^2 \,]$$

$$= 2.4 \times 10^{-6} \text{ N}$$

Example: one gram of Iron

If we had one gram of iron, how hard would it be to remove all the electrons in that sample? From high school chemistry we know that Iron has 55.85 grams per "mole", and the number of moles "n" equals the number of atoms or molecules in the sample divided by Advogadro's number (6.02×10^{23} atoms/mole):

$$n = \text{\#atoms / Advogadro's number}$$

which is also equal to:

$$n = \text{sample mass / (Atomic Mass)}$$

$$= \text{1 gram / (55.85 g/mole)}$$

$$= \quad .018 \text{ moles of Iron}$$

Equating this with the first equation, we find the number of atoms in 1 gram of Iron is:

$$\#atoms = (.018 \text{ moles}) \times 6.02 \times 10^{23} \text{ atoms/mole}$$

$$= 1.08 \times 10^{22} \text{ atoms}$$

Combining this with the results from the previous example, we find the total estimated force required to remove all of these electrons would be:

$$F = (2.4 \times 10^{-6} \text{ N / atom}) \; (1.08 \times 10^{22} \text{ atoms})$$

$$= 2.6 \times 10^{16} \text{ N!}$$

By comparison, it would be far easier to lift a WWII battleship (~65,000 tons or 0.6×10^{9} N) into the air than it would be to extract all of these electrons.

<u>Postulate # 2)</u> <u>Superposition</u>:

In our attempt to model an atom, we find that no simple mathematical function exists that accurately describes the structure of an atom, (such as "Sin(x)", or "Tanh(x)", etc). Consequently, we are forced to construct one of our own.

To do that, Quantum Mechanics borrows a technique developed by Jean Baptist Fourier in the late 1800's to describe traveling waves (discussed in most Classical Physics texts, under the topic of "Fourier Wave Mechanics"). Using this technique we can construct virtually any kind of "well-behaved" waveform we need, no matter how complex (e.g. sawtooth wave, triangular wave, etc. – see the square wave example below) by carefully *superimposing* a number of well-chosen "basis" or "building block" functions, such as "Sin(nx)" in the proper proportions:

$$\text{Waveform} = \Sigma\, a_n \text{Sin}(nx)$$

where "Σ" is the Greek letter "s" telling us to *sum* all of the "a*Sin(nx)" terms, "n" is an integer that ranges from 1 to a maximum number "N" (typically the more accurate model includes a very large number of components), and the "a_n's" are the amplitude coefficients of each "Sin(nx)" basis function we include in our approximation – since we don't typically add the same amount of each basis functions (anymore than you would add the same amount of every ingredient used to build a car, or bake a cake, etc.).

Orthogonality:

The only requirement imposed on our set of basis functions "building blocks" used in our superposition, is that it must form a complete set of "Orthogonal" functions – i.e. all of these basis components are "independent" of each other in such a way that they do not "co-mingle", or affect one another when summed together over one full period.

A common example of two orthogonal functions are Cosine and Sine waves. Since Sine waves and Cosine waves are 90° out of phase with one another (one is at zero when the other is at its peak, etc.), the sum of their products over one period is zero:

$$\int_{-\pi}^{\pi} \mathrm{Sin}(x)\,\mathrm{Cos}(x)\,dx = 0 \qquad (3.1)$$

Fourier used these orthogonal basis functions as "building blocks" to enable him to construct models to represent any "well behaved" wave-like function. Conversely, we could also apply this concept in the opposite direction (i.e. we can apply this same thinking to *decompose* complex signals into their simpler, more manageable components). Using superposition, we are able to analyze very complex signals sent through a variety of propagating media (e.g. E&M waves through the atmosphere, the signal path through a radio, a fiber optic cable, etc.), as well as analyze something as complex and intangible as an atomic or molecular system. In essence, this technique exploits the fact that the manifestation of energy we are trying to analyze (e.g. a voice signal, sound waves, radio waves, an atomic energy "cloud-thing", etc.) is rarely composed of a single component. This approach is so powerful and versatile in fact, that it has been applied to an extremely wide range of topics and disciplines from Quantum Mechanics, to radio antenna design, to seismic building analysis, to advanced radio communication signaling techniques (e.g. CDMA and 4G LTE; see our Cell Phone tutorial on our website "EpiphanyBySteveLee.com"), etc.

To illustrate this technique, we will now apply superposition to construct a simple 100Hz square wave. To do so, we simply sum a series of Sine or Cosine waves (each with "just the right" amplitude, "a_k"):

$$S(\omega) = \sum a_k \sin(k\,\omega\, t) \qquad \text{or}$$

$$= \sum a_k \cos(k\,\omega\, t)$$

(where $\omega = 2\pi F$).

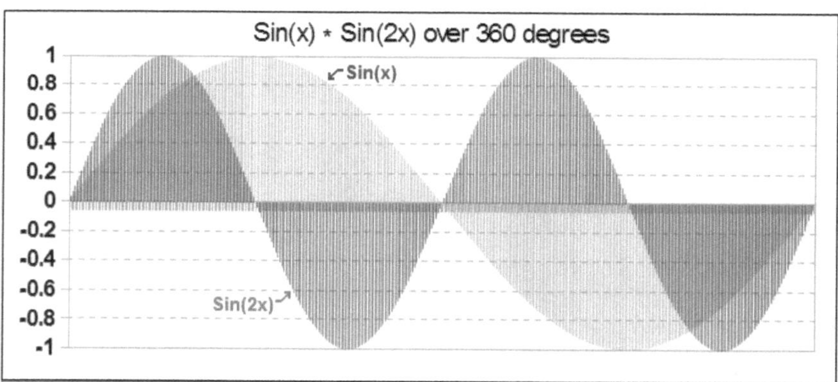

Figure 3.4 The orthogonality of two related Sine waves can be seen by considering the product of areas in this chart. By inspection we can see that the product of the Sin(x) * Sin(2x) alternations produces equal but opposite areas above and below the 0 line. The sum of those equal and opposite areas over the full cycle is 0, indicating Sin(x) and Sin(2x) are orthogonal (as is true for any *integer* multiples of angle "x").

Similarly a Sine wave and discrete multiples of itself represent an Orthogonal set (see Figure 3.4), with "k", "n" and "m" defined to be integers:

$$\int_{-\pi}^{\pi} \text{Sin}(nx)\,\text{Sin}(mx)\,dx \;=\; \pi\,\delta_{(m,n)} \qquad (3.2)$$

where $\delta_{(m,n)}$ (or "Kronecker delta") is defined as:

$$\delta_{(m,n)} = 1 \quad \text{if } n = m,$$

and

$$\delta_{(m,n)} = 0 \quad \text{if } n \neq m.$$

Example: Constructing a Square wave.

By superimposing progressively higher odd multiples of Sin(2 π 100 t) as our "basis set", we can construct a simple square wave (see Appendix B for details). We note that the more terms we include, the more accurately our superposition approaches a true square wave (see figures below):

$$\begin{aligned} S(f) = {} & (4/\pi) * \sin(2\pi\, 100t) + (4/3\pi) * \sin(2\pi\, 300t) + \\ & (4/5\pi) * \sin(2\pi\, 500t) + (4/7\pi) * \sin(2\pi\, 700t) + \\ & (4/9\pi) * \sin(2\pi\, 900t) + (4/11\pi) * \sin(2\pi\, 1100t) + \\ & (4/13\pi) * \sin(2\pi\, 1300t) + (4/15\pi) * \sin(2\pi\, 1500t) + \\ & (4/17\pi) * \sin(2\pi\, 1700t) \;\ldots \end{aligned}$$

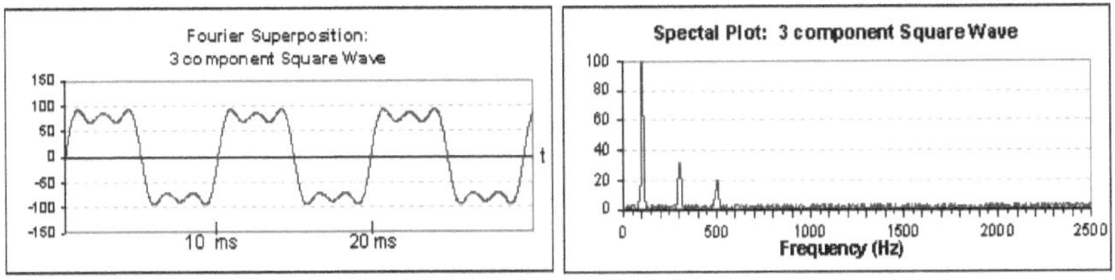

Figure 3.5 An example of Fourier Superposition, generating a square wave using only the first *three* frequency components.

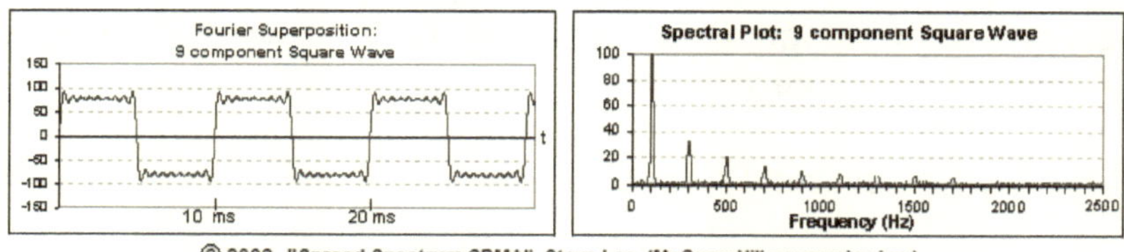

© 2002 "Spread Spectrum CDMA", Steve Lee (McGraw-Hill companies inc.)

Figure 3.6 The same constructed Square wave, using *nine* terms of the superposition.

In the Quantum model of the atom, we exploit the wave nature of mater to allow us to apply this same Fourier wave superposition technique to describe the structure of an atom. In this case, we likewise use a set of orthogonal basis functions ("Φ_n"), which for the time being, we will leave *undefined* (more will be said of this model as we proceed). If we define "$\Psi_{(r,t)}$" to be the overall state of the Atom, then we can write our model as a superposition of orthogonal basis functions "Φ_n's" representing all possible states, using a_n's as the percentage of contribution from each basis function:

$$\Psi_{r,t)} = \sum a_n \Phi_n$$

$$= a_1 \Phi_1 + a_2 \Phi_2 + a_3 \Phi_3 + \ldots \quad \text{etc.}$$

<u>Postulate #3)</u> <u>The de Broglie Wave nature</u>:

As mentioned previously, since atoms and electrons are so tiny, we can't actually see what they look like. However from the Young's double slit experiment, we do know that they are *not* simply hard little spheres with little "b-b"-like electrons "whirling" around them. Furthermore, we know from Einstein's famous $E = mc^2$, that whatever they are, they are *energy.* Therefore, describing them as "nebulous" energy "cloud-things" works as well as anything else.

Being subject to wave-like behavior, we expect that any energy injected into the atom would invoke something of a wave-like undulation behavior vibrating around the atom. This is somwhat analogous to the way kinetic energy "flicked" into a ball of jello would shimmer from the impact point, to the opposite side of the jello ball, and then reflect back to the flick point again (a bit of a stretch perhaps, but you get the general idea). This vibrating energy would continue oscillating back and forth indefinitely around our jello ball if it weren't for the dampening effects of the chunks of fruit and little green marsh mellow bits we all know and love.

Regardless of the actual appearance of the atom, we know the vibrations within the atom are real, they are periodic and hence oscillatory. Therefore, any model we create needs to take this wave-like nature into account. To that end, we will require that whatever we eventually choose for our "Φ_n" basis functions, they must be periodic and wave-like.

Since this point is so central to the Quantum picture, it became necessary to describe all Quantum objects in terms of their "wavelength". In 1923, Louise de Broglie postulated that electrons, protons, neutrons, and hence atoms can be described in terms of a wavelength which is a function of their momentum (and hence their energy), known as the "de Broglie wavelength":

$$\lambda_e = h/p \qquad \text{The de Broglie wavelength}$$

where "h" is Planck's constant, and momentum "p" = mv. By this description, the faster an object moves, the shorter its de Broglie wavelength.

At first glance, this momentum-related wavelength for atomic particles may seem somewhat odd, however there is evidence that support this description:

First, having very specific and very *discrete* radii for each of the hydrogen electron orbits implies discrete circumferences / *path lengths around each orbit.* By analogy, picture each electron orbit around an atom as being something like a tight piano string, only in the shape of a circle (or better yet, a spherical shell). The electrons in this analogy then would be the vibrations traveling around this string/shell that makes up the atomic orbit.

In treating the electron as a wave, we find each orbital radii corresponds to an integer multiple of a half de Broglie wavelength. Any electron that tried to enter into this orbit but had too long or too short a wavelength (i.e. too low or too high an energy), wouldn't "fit" (or "resonate" on our spherical piano string), and therefore would only correspond to an invalid orbit that was "forbidden" in this atom (and consequently would not be absorbed by this atom). This of course corresponds well with what we see empirically in our spectral data.

Another bit of information that aligns with this wave-like picture of an electron is that the faster an electron travels, the shorter its wavelength, and the greater its kinetic energy. This corresponds with our previous statement about light, in that the shorter wavelengths of "light" (e.g. ultraviolet and x-rays) have greater energy than longer wavelength "light" (e.g. infrared and radio waves), as per $E = hf$. This is borne out by the fact that x-rays (for example) penetrate solid objects, while infrared and radio waves do not.

Using this wave-like model (and thus wave-like basis functions as our building blocks) with specific momentum-related wavelengths for electrons, etc., allows us to employ not only the Fourier Superposition technique, but also other Fourier wave mechanic "tools" (as we shall see), which greatly simplify our analysis of this "invisible" realm.

<u>Postulate # 4)</u> <u>The Heisenberg Uncertainty Principle</u>:

The fact that these sub-microscopic objects are so difficult to detect or measure, is further complicated anytime we attempt to simultaneously measure two "mutually exclusive" aspects of such an object, such as position and momentum (where momentum "p" = m*v).

In order to accurately measure an object's position, we need it to remain in a location long enough to get a good sample (i.e. ideally to be at rest). This makes measuring its velocity (or momentum) at the same time very difficult, since to measure velocity, we need the object to remain at a constant speed and direction long enough to acquire a descent velocity measurement. As a result, measuring position (x) and momentum (p) to perfect accuracy (i.e. no uncertainty in either) *simultaneously* of a Quantum wave-like object is impossible.

The "Heisenberg Uncertainty Principle"[12] codifies this concept, by stating that the product of the uncertainty in the position, times the uncertainty in momentum is never less than some minimum value h/2π (where "h" is "Planck's constant" = 6.626×10^{-34} J-sec):

$$\Delta x * \Delta p \geq h / 2\pi \qquad \text{Quantum wave uncertainty}$$

In other words, we can devise an experiment where Δx is very small, but in doing so the uncertainty in momentum (Δp) becomes extremely large (and vice versa). If we manage to somehow find a way to measure the position with *zero* uncertainty, the Heisenberg Uncertainty aspect of the system will result in Δp being *infinitely* large.

The wave-related aspect of this concept was understood well before the advent of Quantum Physics, in that Fourier in his work on classical wave mechanics, expressed the same difficulty when attempting to measure a classical wave moving through some propagating media (e.g. radio waves, light, sound waves, etc.). Waves after all, do not have well-defined edges (compared to say, a billiard ball, or a speeding locomotive, etc.). Consequently there is a certain "ambiguity" to all waves and superimposed wave "packets" especially, since the combination of waves (each traveling with slightly different speeds) inherently adds to that ambiguity. As a consequence, it is harder to know any wave's position and its velocity with zero uncertainty.

$$\Delta x * \Delta p \geq 2\pi \qquad \text{Classical uncertainty limit}$$

---

12 We note that in addition to "x" and "p" there are several other "mutually exclusive" variables, including energy (E=hf) and time (etc.), giving us: $\Delta E \, \Delta t \geq h / 2\pi$. This becomes an issue when we discuss the line width of the laser beam, where we will find that laser beams typically are not as monochromatic as we might like, since this relation forbids $\Delta E \, \Delta t = 0$.

Stepping back from the nuts and bolts of the equations for a moment, and looking at this from a broader perspective, we find this principle reiterates our conclusions in Postulate #1. The general "fuzziness" involved in trying to study Quantum objects inherently implies there will always be some uncertainty in our knowledge about any single Quantum object we may want to study (we will see a bizarre consequence of this uncertainty later when we discuss "tunneling").

All of this of course goes hand-in-hand with our applying Fourier's wave "algebra" to atoms and their sub-components. This in turn underscores our need to bear in mind that electrons, protons, etc. are not only *discrete*, but *wave-like* objects, with no well-defined edges. Consequently there is *always* an element of uncertainty inherent in any system they compose.

<u>Postulate # 5)</u> <u>Duality</u>:

Having said that Quantum objects are wave-like, we now brutally introduce the reader to one of the more unsettling aspects of Quantum Physics: its propensity for an endless array of apparent paradoxes and dualities. Though it is true that we can prove that electrons are *wave*-like (e.g. using the Young's double slit experiment), however we can also demonstrate that they exhibit *particle* behavior as well in certain situations (which Einstein amply illustrated in his thesis on the photoelectric effect). Hence, these objects have something we describe as a "wave-like" / "particle-like" *duality* to them.

An atom, for example, being composed of wave-like electrons and protons, must itself therefore be wave-like. And if we perform Young's double slit experiment using whole atoms (rather than just electrons), we see that they do in fact produce a result that only energy in the form of a wave can generate.

One of the problems in attempting to teach Quantum Physics to students, is that much of what it seems to be telling us almost contradicts what our "common sense" perception of the world has led us to believe ("What, now its *both* a wave *and* a particle?"). Consequently, we tend to resist the Quantum model, since we have such a difficult time relating it to any macroscopic objects around us. After all, when was the last time you saw a billiard ball, or block of wood exhibit "wave-like" qualities?

Part of the problem in grasping this description of nature however, is that our "common sense" perception of things around us is based on our everyday experiences, which in turn depends heavily on our senses (e.g. sight, touch, smell). The flaw is in the fact that our senses were never designed to detect the infinite array of atomic-level detail that underlies the fabric of the universe itself. They were instead designed to analyze things on the *macroscopic* level, effectively "averaging out" all the microscopic details that would otherwise be too overwhelming for our brains to process. Consequently we have over our lifetimes built up a set of perceptions and expectations about the world around us, all based entirely on our "flawed", or at the very least *incomplete* macroscopic view of things.

In other words, our interpretation of our cumulative experiences has over our lifetimes only reinforced a flawed perception of the world around us, and lacking any other information to the contrary, we have been perfectly happy to accept these perceptions. For example, we have come to believe that things such as books, tables, and chairs are all solid objects with well-defined edges. However we know that everything (including ourselves) is composed of atoms, and that each of these atoms is in fact mostly "empty space" (a little "spec" of a nucleus, surrounded by electrons that are at least 1800 times smaller "orbiting" a considerable distance away, by comparison). This tells us that all "solid" objects, *including ourselves*, are composed mostly of empty space.

Since electrons repel other electrons, when we touch one of these "solid" objects, all the electrons at our fingertips repel all the electrons at the surface of that object, and it is this repulsive force that defines what we perceive to be the hard boundary between our fingers and these "solid" objects.

The bottom line in all of this is that *reality is what it is*, and our *flawed* perceptions sometimes leads us to see it in an inaccurate, or at the very least "incomplete" way. In order to grasp what Quantum Physics is trying to tell us, we must be able to accept that our "common sense" perceptions of things is incomplete, and be willing to re-assess what we *think* we know, in order to expand beyond that towards a more accurate understanding of the world around us.

Chapter 3 Review questions:

1. What "trick" did Max Planck use to create an accurate model of Black Body Radiation?
2. What did Planck's solution imply about matter?
3. How did Einstein's work on the photoelectric effect reinforce Planck's contribution?
4. Why did Planck and Einstein not like the new Physics that emerged?
5. What is the requirement on a photon's energy before it can be absorbed by an atom?
6. Why are the atomic energy levels of Atoms and Molecules *discrete*?
7. How does lifting water against the force of gravity store energy in it? How does lifting an electron against the attractive force of the nucleus store energy in it? How are these similar? How are they dissimilar?
8. What is Florescence? Phosphorescence? How are they related to the phenomenon of lasing?
9. What is a "Metastable" state? Why is it important?
10. What are spontaneous emissions?
11. What are stimulated emissions? Why are they important to lasing?
12. Why is lasing so inefficient (name specific loss mechanisms)?
13. Describe the Bohr model of the atom.
14. What is the primary implication of Bohr's model?
15. What hard data supports Bohr's model?
16. What is a Bohr Radius and how does it relate to spectra of ionized gasses?
17. In Figure 3.3, which spectral line corresponds to which level transition (hint: $E=hf = h\ c/\lambda$)?
18. Why do we often refer to the Bohr model as a "stick figure" approximation?

19. Describe Young's Double Slit experiment.
20. What do we learn about electrons, protons, etc. when we apply Young's Double Slit experiment to Quantum systems?
21. Name two reasons why we can't directly observe individual electrons.
22. How do we "contaminate" a Quantum system by merely observing it?
23. What is the principle of Superposition and how do we use it in describing Quantum systems?
24. What is the de Broglie wavelength, and why do we use it to describe Quantum objects?
25. What is the Heisenberg Uncertainty Principle?
26. Name two reasons for imposing the Heisenberg Uncertainty Principle in our study of Quantum objects.
27. What is the principle of Duality, and how does it apply to Quantum systems?
28. What does "Duality" imply about our perceptions of matter?
29. Following the examples in the text, calculate the *fourth* Hydrogen orbital radius ("r_4") and the wavelength of light given off due to a 4==>1 orbital transition ("$\lambda_{4,1}$"). Is this in the visible range?

Chapter 4: Quantum Physics and Lasing

Quantum State numbers:

Thus far we have developed a fairly basic *first approximation* picture of the atom using a blend of classical concepts (e.g. a "ball and spring" model, coupled with the Lorentz Force, and Coulomb's Law). We then modified this picture due to what was discovered by people like Planck, Einstein, and others, who were able to show that the very discrete absorption and emission of energy by gasses implies we must use a *discrete* model of the atom (with well-defined, *quantized* orbits separated by "forbidden" areas in-between), which led us to Bohr's model. This picture of a discrete structure, underscored by strong "wave-like" characteristics, brings us a little closer to understanding the true fabric of nature. In the process of developing this more refined model, we specified each of the *quantized* orbits (or energy levels) around the atom using an energy level number "n", which we will now label the "*Principle Quantum Number*".

When people began comparing the results generated using Bohr's model to the experimental data already available in the form of spectral lines emitted by various glowing gases, they found that it agreed remarkably well with that data, further reinforcing the trend towards the highly discrete Quantum model started by Planck and Einstein.

Under close scrutiny however, several people began to notice that many of the individual spectral lines were often "split", forming a number of very closely spaced fine lines. Since different wavelengths correspond to different energy levels, it is clear that these split lines indicate that each individual "orbit" loosely represented in the Bohr model by "n" must be further subdivided into several discrete, closely spaced "sub-orbitals" – small *perturbations* to the orbital energy, as it were.

These findings suggested slight variations existed around the principle quantum number "n" (i.e. slight variations in each orbit's energy level) that accounted for this fine detail aspect of the spectra data. Soon scientists were attributing these fine detail variations to the motion-induced effects of the electron in its orbit, such as its "angular momentum", its "spin", etc. After all, even classical physics tells us that an object's total energy is a combination of its Potential Energy (e.g. a car parked on the top of a hill, or an electron caught in the electrostatic attraction between it and the protons in the nucleus, etc.) + its Kinetic Energy (the energy represented by its motion), i.e.

$$E_{tot} = PE + KE$$

To account for the energy in the electron's more "subtle" motions that was apparently splitting these orbital energy levels, a refined model of the atom needed to include what became known as the "*Orbital Angular Momentum number*" designated by "L". As in the case of the Principle Quantum Number "n", the Orbital Angular Momentum number can only take on specific, *discrete* values, ranging from 0 to (n-1). Each of these values became known as a "sub-orbital" with a specific letter designation, but since "l" can only range up to (n-1), not all sub-orbitals exist for each orbit (see below). In addition, each suborbital has a maximum number of electrons that it can accommodate:

l = 0: "s" sub-orbital, can hold up to 2 electrons
l = 1: "p" sub-orbital, can hold up to 6 electrons
l = 2: "d" sub-orbital, can hold up to 10 electrons
l = 3: "f" sub-orbital, can hold up to 14 electrons
etc.

Upon even closer scrutiny, experimenters noticed an even finer level of detail existed that further split even these sub-orbitals. To account for these finer perturbations, two more subdivisions were required including the "*Magnetic Quantum number*" and finally the "*Spin Quantum number*" (i.e. suggesting the electron is in one of two "spin" states: spin-up "+½" -or- spin-down "-½"), each with a very well-defined, discrete range of possible values.

Shell (n) ↓ / sub-shell (l) →	s	p	d	f	g				
1	2					=	s	=	2
2	2	6				=	s + p	=	8
3	2	6	10			=	s + p + d	=	18
4	2	6	10	14		=	s + p + d + f	=	32
5	2	6	10	14	18	=	s + p + d + f + g	=	50

Figure 4.1 Atomic orbits (n) and sub-orbits (l), with electrons distributed through the sub-orbitals of each shell as shown.

Table 4.1

Symbol	Name	Range of values
n	Principle Quantum number	1, 2, 3, ... ∞
l	Angular Momentum number	0, 1, ... (n-1)
m	Magnetic Quantum number	-l, (-l+1) ... 0 ... (l+1), l
m_s	Spin Quantum number	-1/2 or +1/2

As a result of these various possible sub-states, there are a total of n^2 possible "vacancies" or states that can be occupied by electrons per atom. For example, Carbon with six protons, attracts six electrons: two fill the first orbit ($1s^2$), n=1 spin up, and n=1 spin down. Of the remaining four, two then go into the n=2 "s" sub orbital (one as spin up and the other as spin down), and the remaining two then go into the n=2 "p" sub orbital (one as spin up and the other as spin down).

Why only one electron in each state? Because "like charges repel each other", forcing any two near each other to separate. This concept is codified in a rule that became known as the "*Paulie Exclusion Principle*", which states that each vacancy can be occupied by *one* and *only one* electron. This exclusionary occupation, causes the electrons to "fill" an atom in a somewhat ordered and predictable way, beginning with the lower energy levels and progressing to the higher levels, as shown in the following table of atomic electronic configurations:

Table 4.2 Electronic configurations and the Periodic Table[13].

1	Hydrogen	H	$1s^1$				
2	Helium	He	$1s^2$				
3	Lithium	Li	$1s^2$	$2s^1$			
4	Beryllium	Be	$1s^2$	$2s^2$			
5	Boron	B	$1s^2$	$2s^2\ 2p^1$			
6	Carbon	C	$1s^2$	$2s^2\ 2p^2$			
7	Nitrogen	N	$1s^2$	$2s^2\ 2p^3$			
8	Oxygen	O	$1s^2$	$2s^2\ 2p^4$			
9	Fluorine	F	$1s^2$	$2s^2\ 2p^5$			
10	Neon	Ne	$1s^2$	$2s^2\ 2p^6$			
11	Sodium	Na	$1s^2$	$2s^2\ 2p^6$	$3s^1$		
12	Magnesium	Mg	$1s^2$	$2s^2\ 2p^6$	$3s^2$		
13	Aluminum	Al	$1s^2$	$2s^2\ 2p^6$	$3s^2$	$3p^1$	
14	Silicon	Si	$1s^2$	$2s^2\ 2p^6$	$3s^2$	$3p^2$	
15	Phosphor	P	$1s^2$	$2s^2\ 2p^6$	$3s^2$	$3p^3$	
16	Sulfur	S	$1s^2$	$2s^2\ 2p^6$	$3s^2$	$3p^4$	
17	Chlorine	Cl	$1s^2$	$2s^2\ 2p^6$	$3s^2$	$3p^5$	
18	Argon	Ar	$1s^2$	$2s^2\ 2p^6$	$3s^2$	$3p^6$	
36	Krypton	Kr	"Argon" + $3d^{10}$		$4s^2$	$4p^6$	etc.

Exercise:

Using this approach, complete the electronic configurations for the elements between Argon and Krypton in the Periodic Table of Elements.

Why do we care how electrons distribute around an atom? Because the distribution of electrons across each atom's defined orbits determines the number of electrons that end up in the outer-most orbit, and it is this outer orbit that determines how each element interacts with all other atoms. In other words, this very specific distribution of electrons is what gives each element

---

[13] Note: this sequence is only an approximation. In reality, larger atoms do begin to place electrons in some of the next higher sub-orbitals prior to completely filling the current sub-orbital.

its characteristic chemical properties, including their bond strengths, which in turn determines the amount of energy needed to excite each bond into oscillation (which in turn defines its "spectral signature").

For example, Hydrogen has only one isolated electron in its outer shell, which is the 1s shell (recall that s-subshells can hold a maximum of *two* electrons). Consequently Hydrogen tends to be "incomplete"[14], and hence is eager to pair with other atoms looking for additional electrons to complete their outer shells. Oxygen on the other hand has only *four* of the total *six* possible electrons in the 2p subshell, and it therefore needs two more electrons to complete its outer shell. As a consequence, when Oxygen and Hydrogen encounter each other, the Oxygen eagerly bonds with two Hydrogen atoms to form H_2O (water).

Chlorine lacks only one electron to give it the desired eight electrons in its outer shell, and as a result eagerly bonds with Sodium which has but one isolated electron dangling out there all alone in the 3s shell. As a result, these two atoms are very eager to bond together, typically in a *violent* reaction as they form NaCl (common table salt).

Helium and Neon, on the other hand both have completed outer shells and hence have absolutely no interest in bonding with any other atoms. As a result, these elements tend to be very non-reactive chemically, and hence are rarely ever found in nature as anything other than isolated atoms. Such elements are known as "*Nobel Gases*".

---

[14] We note that the "magic number" for completeness is *two* for the *first* shell (n=1, e.g. Hydrogen and Helium) and *eight* for the *second* shell (e.g. Carbon or Oxygen). For n=1, "l" can only go as high as 0, hence only the "s" sub-shell exits in the first shell. Since an "s" subshell can only accept two electrons, *two* is a "magic number". For n=2, "l" ranges from 0 to 1 (i.e. "s" and "p" sub-orbitals). With both the "s" and the "p" sub-orbitals present, a total of 2+6 = 8 electrons can be accommodated, hence the "magic number" *eight*.

Neon	:Ne:	H_2O	H·:O:·H
NaCl	Na.·Cl:	CO_2	:O::C::O:

Figure 4.2 Molecular bonding

Now that we understand the basic "rules" governing the structure of the atom and hence its chemical interactions, we can predict the likelihood two atoms will bond to form a molecule, giving us an idea of the strength of any such bond. Such insights can be very useful as we design our lasers.

From our previous discussions about Carnot engines and Thermodynamics in general, it is clear that all things in nature seek to be in a more stable, less energetic configuration, be they boulders on mountaintops, or electrons in an elevated shell around a vacancy in an atom, etc. As with the macroscopic world, we find that this tendency to seek microscopic stability can also be exploited to perform some useful work in the process. In fact, this is the most fundamental concept behind virtually all of our modern technologies, including electronics, chemistry, biomedicines, nuclear physics, lasers, etc. Knowing the preferred state and behavior of these systems, we exploit that behavior by providing them a means to achieve that preferred state, having them perform some useful task for us along the way.

Lasers are a prime example of this, in that we could not possibly construct the *trillions* of devices needed to "spin" our raw energy into the "gold" of coherent monochromatic light. Instead we identify atoms/molecules that naturally possess the characteristics we seek, and have them put those characteristics to work for us. In this process, we let them do what they do naturally along some prescribed path that just happens to channel their actions to assist us in our quest towards some goal (e.g. to amplify a weak signal through a transistor, or to feed into a metastable state to enhance a population inversion, etc.).

We have even gone so far as to use this approach to exploit some of the fundamental properties of Noble Gases to produce a unique class of lasers known as *"Excimer lasers".* The term "Eximer" is a contraction of "Excited Dimer", with "di-mer" being a polymer with only two ("di") molecules ("mers"). In

Excimer lasers, we artificially cause a Nobel gas (e.g. Krypton) to behave much the same way Sodium does, by exciting the atom (in a multi-step process described below, including bombarding it with a beam of electrons). Once placed in this excited state, the "Nobel Ion" ends up with one electron in its outer-most orbit (similar to Sodium). In this state, it then eagerly bonds with a Halogen (e.g. Fluorine). Shortly thereafter, the noble gas de-excites, causing the Eximer to discard the Halogen the instant it re-stabilizes and once again becomes a true Nobel gas. In the process, it violently expels the Halogen with a tremendous amount of energy in order to return to its more natural isolated configuration.

One of the more common examples of Excimer lasers in use today is the UltraViolet KrF laser (248 nm), which is pumped according to the following chemical reactions. We initiate the process by bombarding Argon, Fluorine and Krypton with an electron beam, which creates the ions shown below, plus a few surplus electrons which promote the separation of the Fluorine molecules F_2 into separate Fluorine ions. A high concentration of Argon (~85-90%) is used in the gain media with the primary function of promoting the creation of Krypton ions:

$$Ar^+ + 2Ar \Rightarrow Ar_2^+ + Ar$$

$$Ar_2^+ + Kr \Rightarrow Kr^+ + 2Ar$$

$$F_2 + e^- \Rightarrow F^- + F$$

$$Kr^+ + F \Rightarrow KrF^*$$

(where we used " * " to indicate an excited state.)

De-excitation typically follows very rapidly (< 10^{-12} seconds), as the two atoms undergo a violent "divorce". In so doing, they release a fair amount of energy in the form of coherent monochromatic light.

Another example of the exploitation of the natural preferences of atoms and molecules is the class of lasers known as "chemical lasers". Chemical lasers rely on the fact that when certain chemicals are mixed together, they create an *exothermic* reaction (i.e. a reaction that expels energy, typically as heat). In the case of Chemical lasers however, the energy released is not only in the form of heat, but also in the form of photons (much the same way "Glow Sticks" release visible light when activated by rupturing the barrier between the two reservoirs containing the agents involved in their reactions).

Examples of Chemical lasers include the Hydrogen-Fluoride (HF) laser, the Deuterium-Fluoride (DF) laser, and the COIL (Chemical Oxygen-Iodine laser). This latter laser now has output powers in the neighborhood of 20 kW or better, around 2-3 μm, generated by mixing Hydrogen Peroxide and Potassium Hydroxide with Chlorine in a very exothermic reaction (>100 kJ/mol) [20].

Selection Rules and the *Metastable state*:

When we first described lasing, we mentioned that the gain media must be excited into a metastable state in preparation for stimulated emissions. Once energy is channeled into a metastable state, it is held there for an abnormally long period of time as a result of the fact that the atom or molecule in the gain media is unable to de-excite to its ground state. To explain this behavior, we must consider the probability of transition between the excited state and the ground state. This analysis leads us to what are known as "*Dipole Selection Rules*" (or simply "Selection Rules", see Appendix H).

When we discussed the various possible "orbital" states, we found that there were certain specific restrictions as to what orbits were allowed and which ones were forbidden, due to the very *discrete* amount of energy present in the system. This led to very well-defined values for the orbital radii, as well as a range on the other allowed values for each of the Quantum numbers. For example, the Principle Quantum number "n" can only take on integer values, which when coupled with the previously developed equations pertaining to the Bohr radius, specified very clearly defined allowable states, with all other orbits in-between being "forbidden". The same is true for the Angular Momentum sub-orbital number "l" which can only assume a value from 0 up to "n-1", and the electron Spin number "m_s" which can only be either "+½" or "–½". When we analyze the *transition* between any possible states, we find one more very rigid limitation applies:

When an electron changes orbits, it *must* also change sub-orbitals at the same time. In other words, anytime an electron changes orbits (i.e. when "n" changes), "l" *must* also change by *one* (i.e. it has to go into a sub-shell in the new orbit position *different* from the sub-shell it was in). This Selection Rule is a consequence of the Pauli Exclusion Principle which forbids an overlap between two states in such a dipole configuration (see Appendix H for all the ugly mathematical details).

For example, if an electron were at n=2 and l=0 (the "s" sub-orbital in the second shell), and it wanted to fall in to fill a recent vacancy in the first shell (n=1), the Dipole Selection Rule forbids it. This transition is "forbidden" since the first shell only has an "s" sub-orbital, and according to our Selection Rule, "l" must change by *one*. Consequently this electron will remain inverted in this excited state for a *prolonged* period of time (e.g. ~tens of milliseconds for florescent materials, and seconds or longer for phosphorescent materials):

$\Delta l = \pm 1$ Selection Rule for sub-shells[15]

This is the origin of what we have been calling the "*Metastable state*", and it is this purely Quantum Mechanical mechanism that makes lasing possible. Without this restriction, anytime any atom or molecule became excited by absorbing energy from some outside source, it would *immediately* de-excite (typically within ~10^{-15} seconds) by falling back in to whatever vacancy was available, and hence no population inversion could be setup to enable any subsequent stimulated emissions.

15 In Chapter 5 we will find that in addition to the Δl rule, there is another similar "Selection Rule" (ΔJ) for molecules that relates to angular momentums.

Without this mechanism, lasing and hence lasers would not be possible.

<u>Schrödinger Equation</u>:

At this point we are about as ready as we are going to be to introduce what is the arguably the most pivotal mathematical equation to emerge from the field of Quantum Physics, namely the *Schrödinger Equation*. Though we will offer a weak stab at *appearing* to derive it from basic principles, it actually has no honest, *legitimate* derivation. Like de Broglie, Bohr and others, Erwin Schrödinger in 1925, basically took some general classical concepts (e.g. the electromagnetic wave equation) and arbitrarily threw in a few Planck's constants and a couple complex numbers, and *voila*: out pops one of the most profound and fundamental discoveries of the 20th century (and we wonder why Einstein was so unnerved by this vague, hand-wavy, "trailer-park" version of his beloved Physics. *Plastic hubcap wall ornament anyone*?).

However unnerving its origins, the best justification available for Schrödinger's Equation, is that *it works*. In fact we have well over a century of spectroscopic data which bears this out remarkably well (not to mention some very interesting results from rather exotic disciplines such as the study of sub-atomic particles and Quantum field theory, as well as molecular spectroscopy, semiconductor physics, and lasers themselves). Beyond the hard tangible data, there is also something of a "warm-fuzzy feeling" we get when we notice it has some *vague* similarity to the Classical wave equation, which *is* derivable (see our brief tutorial on Maxwell's Equations, at "EpiphanyBySteveLee.com", misc. tab):

$$d^2 \mathbf{E}/dx^2 - (\varepsilon \, \mu \, /c) \, d^2 \mathbf{E}/dt^2 = \text{source}$$

(The Classical E&M wave equation)

Since we can gain some very useful *conceptual* insights by going through the motions of *pretending* to derive Schrödinger's equations from basic concepts, we will do so here:

Using $\Psi_{(x,t)}$ as the "State Function" describing the energy state of our Quantum system, "H" as the "Hamiltonian Energy operator" = Kinetic Energy + Potential Energy (H = K + U), and "E" as the energy state of the system (see previous discussion on atomic orbitals and sub-orbitals, etc.), we begin our "*derivation*" (*I can't even say it with a straight face*):

As an "Operator", anytime we apply the "Hamiltonian Energy Operator" (H) on our state function, it produces an output, in this case the energy level (E) the state is currently in. So, applying "H" to the state function:

$$H \Psi_{(x,t)} = E \Psi_{(x,t)}$$
$$(K + U) \Psi_{(x,t)} = E \Psi_{x,t)}$$

From Classical physics, we know that Kinetic Energy ($KE = ½ mv^2$) can be written in terms of momentum ($p = mv$): $KE = P^2 / 2m$. If we define "P" to be the "Momentum Operator" with the [*arbitrary*] definition of: $P = -i \hbar\ d/dx$ and [*arbitrarily*] define "E" to be the following operator: $i \hbar\ d/dt$ (with $\hbar = h/2\pi$), plug this into our equation (*while stroking a few crystal trinkets purchased from the Area 51 souvenir shop*), we get:

$$(P^2 / 2m + U) \Psi_{(x,t)} = i \hbar\ d\Psi_{(x,t)} / dt$$

$$([-i \hbar\ d/dx]^2 / (2m) + U) \Psi_{(x,t)} = i \hbar\ d\Psi_{(x,t)} / dt$$

$$(-\hbar^2 / (2m)\ \ d^2/dx^2 + U) \Psi_{(x,t)} = i \hbar\ d\Psi_{(x,t)} / dt$$

Distributing our state function throughout, our equation becomes:

$$-\hbar^2 / (2m)\ \ d^2\Psi_{(x,t)} /dx^2 + U \Psi_{(x,t)} = i \hbar\ d \Psi_{(x,t)} / dt$$

which is *Schrödinger's equation.*

Though at first glance this result may appear somewhat hideous, if we remind ourselves that "U" is simply the potential energy stored in the system (e.g. the potential energy stored in the Coulombic attraction between nucleus and the negatively charged electron: $- e^2 / [4 \pi \varepsilon r^2]$, or stored in the restoring force of the molecular bond "springs", ½ kx^2), and that the other terms (e.g. "h", "m" and "i") are "*nothing more than*" constants, we can then almost begin to convince ourselves that it bears at least some passing similarity to the Classical Wave equation (shown above). *Sort of.*

At the very least, it should be apparent from being forced to endure the above hand-wavy *Vaudeville act,* that this equation *is a statement about the energy in the system:* The left hand side of the equation represents the total energy (Kinetic + Potential), while the right-hand side is the resulting change of the energy in the system with time. In other words, by applying our Hamiltonian energy operator on the state function on the left, it then spits out the specific level of energy currently in the system on the right hand side (just like that nifty Play-Dooh extrusion toy you loved so much as a kid... and occasionally pull out of the closet when no one is looking).

Once again, for those who are still somewhat unconvinced, we stress that the real validation of Schrödinger's equation is that *it works*, as we shall soon see. Whenever it has been applied to the state functions of an atomic or molecular system (e.g. see Appendix G), the results it generates agree remarkably well with actual scientific data (e.g. gas spectral plots, etc.). Even though it may not look all that *warm and fuzzy*, and its *terrestrial* lineage somewhat "*dubious*", its ability to solve problems otherwise unsolvable through any other means, has earned it a high degree of respect in the field. To that end, we now offer a few examples to show Schrödinger's equation applied to some *simple* Quantum systems:

Example: Free electron

Consider an electron operating in free space, with no external field or "restoring" potential (i.e. U = 0). Then Schrödinger's equation simplifies to:

$$-\hbar^2 / (2m) \;\; d^2\Psi_{(x,t)} / dx^2 \;+\; 0 \;=\; i\hbar \;\; d\,\Psi_{(x,t)} / dt$$

Assuming a wave-like description for our free space electron of the following form:

$$\Psi_{(x,t)} = A \exp(ikx - i\omega t)$$

we insert this into Schrödinger's equation above to produce:

$$-\hbar^2/(2m) \quad d^2 A \exp(ikx - i\omega t)/dx^2 = i\hbar \; d \; A \exp(ikx - i\omega t)/dt$$

Performing the differentiation, we get:

$$-(i)^2 \hbar^2/(2m) \; k^2 A \exp(kx - \omega t) = -i\hbar \; i\omega \; A \exp(kx - \omega t)$$

Note that when we use our Quantum Operators on our state function, that the energy is extracted from the function and the original state function remains. This is a fundamental aspect of Operators, and of Quantum Mechanical operators in particular (see discussion on Eigenstates below). Isolating our original equation terms to identify the change, we find:

$$\hbar^2/(2m) \; k^2 \Psi_{(x,t)} = \hbar\omega \; \Psi_{(x,t)}$$

We can further isolate the result:

$$\hbar^2 k^2/(2m) = \hbar\omega$$

Recall that we previously mentioned that the energy in a photon is directly related to the wavelength or frequency, i.e.: $E = hf = \hbar\omega$. If the charged particle is in an area where no electric field exists (i.e. it is a "free" particle), it has no potential energy (U=0). Therefore its total energy simply equals its kinetic energy ($p^2/2m$). Combining these results, we get:

$$E = \hbar^2 k^2/(2m) = p^2/2m$$

Hence:

$$p = \hbar k, \qquad \text{or} \qquad k = p / \hbar$$

From Classical E&M, the definition of "k" $= 2\pi / \lambda$. Therefore:

$$k = p / \hbar = 2\pi / \lambda$$

or:

$$\lambda = 2\pi\hbar / p = h / p$$

Which is simply the equation for the de Broglie's wavelength for our free electron. So far, so good.

<u>Eigen-States, Eigen-Operators, and Eigen-values</u>:

In the above example we noted that when we operated on a state function using our Hamiltonian Energy operator, it generated a discrete value and returned our original state function. Many of the equations and operators used in Quantum Physics fall into a special class, which have this discrete "input", "output" relationship. These equations are known as "Eigenvalue[16] Equations", with the operators used in them being known as "Eigenoperators" and the results they generate being therefore known as "Eigenvalues". If two different states have the same Eigenvalue output, they are said to be "*Degenerate*", and as a result are *indistinguishable* from one another via any measurement we can make (e.g. spectral data).

One example of this degenerate behavior is found in the vibrational modes of the CO_2 molecule (see Figure 2.2), in which the "Y" and "Z" bending modes are indistinguishable in spectral data (they involve essentially the same type of bending [except

---

16 "Eigen" is German for "Characteristic"; hence these equations are used to describe the *characteristically* discrete behaviors of these Quantum systems, which generate predictable and *characteristic* output values.

along different axes], and thus store the same amount of energy when excited).

A complete set of all the Eigenvalues associated with an Eigensystem forms the unique characteristic fingerprint of that system as an *Energy Spectrum.* It is this characteristic fingerprint that allows us to analyze the light given off by a substance or gas to determine its exact composition (e.g. light emanating from stars[17] thousands of light years away can easily be studied in this way to reveal their composition, and hence age).

Stationary States:

A common technique used in solving differential equations, known as "separation of variables", involves breaking the equation into separate and distinct (orthogonal) pieces. This effectively allows us to break a single complex problem into several smaller (less complex) problems. For example, if we were to examine the trajectory of a baseball that has been

---

17 It is believed the life cycle of a star begins with Hydrogen, which it fuses to convert into Helium by virtue of the extreme gravitational pressures created at the center of the star. Once all the Hydrogen is consumed, the fusion reaction dies, removing the outward pressure created by the escaping energy from the reaction. As a result, the star collapses. If the initial mass of the star is great enough to create enough pressure to overcome the Coulombic repulsion between two Helium atoms, the star will begin to fuse Helium into larger atoms, including Carbon, creating a new pressure equilibrium in the process. Normal sized stars ("main sequence" stars) such as our own sun, eventually fuse all the lighter atoms into Carbon ("Red Giant" phase), but are not large enough to compress Carbon to the point of fusion, and thus end up as White Dwarfs. Larger stars (more that ~4x the size of our sun, "Red Super Giants") do have enough mass to compress Carbon to fuse into progressively larger sized atoms, including iron. Eventually, however, these Super Giants are no longer able to fuse these larger atoms, and once they exhaust all burnable fuel, they collapse in a supernova. The resulting internal pressures are so extreme that they compress much of these larger atoms into super heavy atoms (e.g. Uranium). It is believed this is the source of all larger atoms found in the universe.

clobbered by a Louisville slugger (or a cannon ball shot from a cannon, etc.), we would find the easiest approach would be to separate its motion along the "x" and "y" axes in a "divide and conquer" simplification. We can separate these two components because the motion along the "x" axis is independent (orthogonal) from the motion along the "y" axis.

In our Quantum system, $\Psi_{(r,t)}$ is a function of space and time. Treating these two "spaces" as orthogonal spaces, we separate them by writing our state function as a product of a spatial function and a temporal function:

$$\Psi_{(r,t)} = \Phi_{(r)}\ Ш_{(t)}$$

Since we know that our electron is wave-like, we will assume a periodic form for "$Ш_{(t)}$" and thus our state function becomes:

$$\Psi_{(r,t)} = \Phi_{(r)}\ e^{-i\,\omega\,t}$$

Having written it this way, we will now separate our Schrödinger equation into what is known as the "Time dependent Schrödinger Equation" and the "Time independent Schrödinger Equation". This latter form will allow us to consider the simple case of a particle in a stationary state (i.e. not changing in time, analogous to a standing wave on a rope or a transmission line).

Example: an Electron trapped in a square Coulombic well

Consider our electron now trapped within a simple electrostatic well, or box, with walls formed by some negative potential barrier (–V). Inside this box, we will assume there is absolutely no voltage (i.e. V=0). What do we expect the electron to do?

$U(r) = -V$ at "r" = +/- "a",

and

$U(r) = 0$ in between

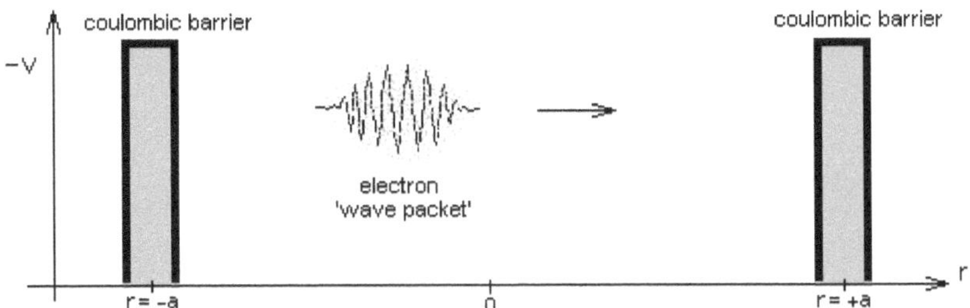

Figure 4.3 Depiction of a quantum particle trapped in a 2-dimensional "box".

Since "like charges repel", we expect the electron to be repulsed by the negatively charged walls. As a result it will tend to remain confined within the potential well formed by the box, establishing a stationary state within this area. Since this box defines two separate regions, we will write a separate equation for each:

$$-\hbar^2/(2m)\ d^2\Phi_{(x)}/dx^2 + 0 \quad = \quad i\hbar E \quad \text{(Region 1, in box)}$$

$$-\hbar^2/(2m)\ d^2\Phi_{(x)}/dx^2 - V\,\Phi_{(x)} \quad = \quad i\hbar\ E \quad \text{(Region 2, in walls)}$$

The solution to the first equation is the same as in our "free particle" example, since the conditions are the same (i.e. 0 potential). To solve our equation for region 2, we first rewrite it for simplicity:

$$[\ d^2/dx^2 + (2m)/\hbar^2\ (E-V)\ ]\ \Phi_{(x)} = 0$$

Once again, we note the similarity between this and our Classical wave equation, and try a solution that mimics the solution for a classical electromagnetic wave of the form $e^{i\,\alpha x}$:

$$\Phi_{(x)} \quad = \quad \exp(\ i\ \text{SQRT}\,[(2m)/\hbar^2\ (E-V)\ ]\ x\ )$$

If we plug this back into our equation and grind out the derivatives, we find this function does in fact work, indicating it is a possible solution.

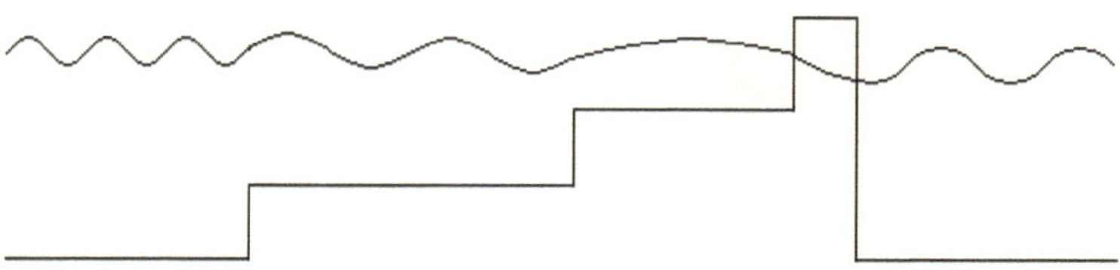

Figure 4.4 An electron wave over various energy regions, and through a barrier. Notice that the frequency (energy) of the electron wave decreases as its height above the potential decreases (i.e. its potential energy decreases). Note also that as it travels through the barrier at the right, it is no longer oscillating but is instead decaying, i.e. being *attenuated* by the barrier (analogous to light passing through glass). When it emerges from the barrier at the far right, it begins oscillating again, but with lower energy (lower frequency).

We note that as long as $E > V$ (i.e. the energy of our electron is *greater* than the potential barrier, e.g. inside the zero potential areas of our "box"), our state function $\Phi_{(x)}$ remains "wave-like". However, when the electron energy is below the level of the potential barrier ($E < V$, i.e. inside the barriers), the square root of a negative number is imaginary. When we combine that imaginary number with the "i" already in the exponential, they cancel giving us $e^{-\alpha x}$, which is no longer sinusoidal but is instead an exponentially *decaying* function (as shown in Figure 4.4 at the right, as the electron wave passes through the barrier "wall", losing energy in the process).

In other words, as long as the energy in the electron is greater than the potential "V", the electron's state function remains similar to what it was as a free particle (though it appears to have a slightly different frequency, if we compare terms between this and the results from our first example). Therefore we see that when $E < V$, the electron's state function "dies off" as it penetrates into the barrier.

Exercise:

Using the results from the previous free particle example ("E_{free}"), show that the energy of the particle over the barrier ("E_B") is simply:

$$E_B = E_{free} - V$$

In other words, the presence of the potential shifts the particle's energy and hence, its frequency (since $E = hf$), as shown in Figure 4.4.

Tunneling:

The results from the previous example beg the questions: What happens when the electron energy is lower than the potential ($E < V$) and the wall is infinitely thick? And what happens if the barrier is relatively thin? Without churning through any more equations, a qualitative answer should be fairly straightforward, noting from the previous information that the wave function decays as it travels through the barrier.

In the case of an infinitely thick wall, the state function continues to decay until it eventually dies out completely:

$$e^{-\infty} \Rightarrow 0$$

In the case of the thin wall, we expect that the function will decay slightly while traveling through the thin wall barrier. Once it emerges from the barrier, and returns to a potential-free zone, it should return to the "free-state" condition and propagate on from there as before, only with slightly less energy (longer wavelength), due to having lost energy moving through the wall.

What we just described in this latter case, is known as "*tunneling*" and can only be fully understood using the Quantum model. In fact, the classical description tell us that if we shoot a particle at a solid wall, it will not "*ghost*" its way through it to emerge on the other side unscathed, but will instead ricochet off. However, there is a great deal of hard evidence that clearly shows that electron tunneling does occur at the atomic level. In fact, electron tunneling forms the basis of the latest development in high resolution, high power microscopy, including what is known as "Scanning Tunneling Microscopy" (STM).

Current STM systems produce some of the highest resolution images of nano-scale samples yet developed, not only enabling us to map the surface of such samples in high resolution, but even going so far as to manipulate atoms to form single atom-wide structures (e.g. IBM used this technique to create a molecular-sized billboard of their logo in 1990 [17]; see Figure 4.5).

In reality the STM does not actually "see" the surface of samples as much as it "feels" it, by repeatedly rastering a charged probe just above the surface of the sample while measuring the change in the current drawn off the surface as a function of position. This current is produced as electrons within the sample tunnel out of the surface and through the space above it into the probe. Since the amount of tunneling current depends on the composition of the sample, we find STM imaging is most effective on samples that are at least partially conductive, and not very effective (or somewhat misleading) for poorly conductive samples.

One of the odd consequences of this "tunneling" effect is that Quantum objects have a tendency to "tunnel" out of what we would classically consider to be a confined space. This of course contributes to a greater uncertainty as to their location and momentum (see our previous discussion on the Uncertainty Principle). Such a "now you see it, now you don't" behavior seems to run counter to our "common sense" expectations, after all, anytime we toss objects into a box, they tend to stay within the walls of that box.

Title: The Beginning Media: Xenon on Nickle (110)

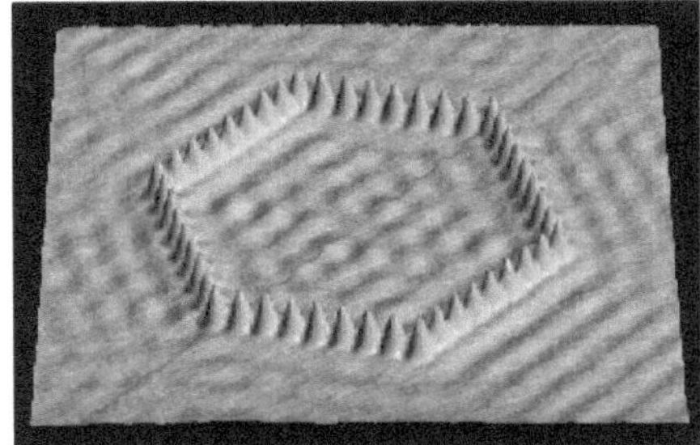

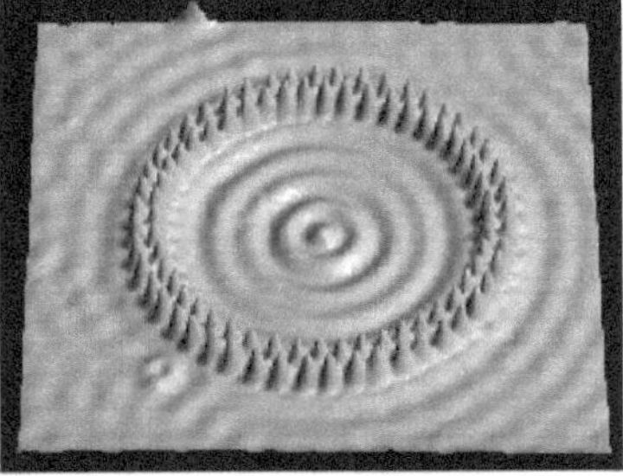

Title: The Corral Collage Media: Iron on Copper (111)

(Images courtesy of IBM Research, Almaden Research Center)

Figure 4.5 IBM STM images (height exaggerated for clarity).

The problem with that macroscopic perception is that Quantum objects (e.g. electrons, atoms, etc.) are not hard little spheres, but are in fact "wave-like" bundles of energy. And we know from living next door to that crazy metal-head college kid for two *painfully* long years, that wave energy (e.g. sound waves, etc.) most definitely can move through walls.

One of the oft-cited consequence of this Uncertainty/"tunneling" effect is what sci-fi buffs like to call instant "sub-space" communications. This behavior tells us that electrons (for example) have a non-zero probability of tunneling through any confining barrier, including the fabric of space itself. Consequently such a behavior suggests we could ("theoretically") communicate *instantaneously* across the vastness of space with "ease" (they do it all the time on *Star*

Gate Atlantis [using a radio no bigger than a *pen* no less], so it *must* be possible).

Quantum Probabilities:

Up to this point in our discussions, we have managed to define our electron's "state function" as a superposition of individual orthogonal wave-like basis functions, whose state in space and time can be described using Schrödinger's equation.

We noted in our first Quantum Postulate that the complications of detecting and measuring something the size of an electron *without disturbing it* in the process is enormous. As a result, seemingly simple questions such as "where is it" or "what is it doing" become almost impossible to answer on an individual case-by-case basis. Fortunately, in most of our applications, we are dealing with very large numbers of atoms and/or electrons, and as a result we do not need to know exactly where each single electron or a single photon is, or what it is doing. In these large number applications, we are quite happy to know only the *likelihood* of an outcome, based on *Statistical Probability* and the *bulk* behavior of the *majority* of other atoms or photons (etc.).

For example, if you have a million Copper b-b's and 10 Steel b-b's of roughly the same size and mass all mixed together, what is the chance of randomly picking one of the steel b-b's out of the pile? 10/1,000,010 or one chance out of 100,000 (rounded). If on the other hand we have half a million of each, our odds of getting a steel b-b would be 500,000 / 1,000,000 = 50%.

Exercise:

What is the chance of getting any "King" out of a normal (well-shuffled) deck of 52 cards? (4/52 = 7.7%)

Exercise:

What is the chance of getting a "5" while rolling a normal (six-sided) die? (1/6 = 16.7%)

Exercise:

What is the chance of getting "three of a kind" rolling three normal (six-sided) dice? (0.463%)

One of the motives for defining our state function as we did, is that it allows us to calculate the *probability* of the object being found in a particular state, by defining the "probability density" $\rho_{(x,t)}$, as:

$$\rho_{(r,t)} \quad = \quad | \Psi_{(r,t)} |^2$$

(which is analogous to the energy of an electromagnetic wave, which is $| \mathbf{E}(r) |^2$). The incremental probability $d\, P_{(r,t)}$ of finding one particle in a small volume "dv" is:

$$d\, P_{(r,t)} \quad = \quad \rho_{(r,t)}\, dv$$

therefore, the total probability "P" is the integral over the whole volume:

$$P \;=\; \int d\, P_{(r,t)}\;\; dv \;=\; \int \rho_{(r,t)}\, dv$$

In other words, the probability of finding a particle depends on where we look (in the volume), and how much of the volume we examine. Of course if we look everywhere (i.e. over an infinite volume), the total probability should be 100% (e.g if you are looking for the ace of spades and you look at *every* card in the deck you have a 100% chance of finding it).

Since our Quantum state functions are represented using complex numbers, when we square the state function we multiply "$\Psi_{(r,t)}$" by its complex conjugate (indicated by the asterisk). To create the complex conjugate of a function, we simply change the sign in front of the "i's" :

$$P \quad = \int \Psi^*_{(r,t)}\; \Psi_{(r,t)}\;\; dv$$

Example:

What is the probability of finding our free electron if we search the whole universe (infinite volume):

$$P = \int A\exp(-ikx + i\omega t) \quad A\exp(ikx - i\omega t) \quad dv$$

$$= A^2 \int \exp(-ikx + i\omega t \; + \; ikx - i\omega t) \; dv$$

$$= A^2 \exp(0) \int dv$$

$$= A^2 \; 4\pi \quad \text{(where } 4\pi \text{ is the volume of a sphere)}$$

Since Probability is typically based on a 0-100% scale (e.g. a 400% probability of finding something is a little odd), we typically "normalize" the state functions (i.e. we divide by the maximum) to insure a 100% probability anytime we examine the full range or volume. Thus to normalize the state functions in the previous example, we define "A" to be: $A = 1/\sqrt{(4\pi)}$.

<u>The "Soliton" wave packet</u>:

Before winding down this overview of Quantum modeling of electrons, etc., there is one more classical topic worth mentioning, since it aligns well with our description of electrons as "wave-packets". That topic is the classical study of "Soliton" waves.

Solitons are effectively "bundles" of waves traveling together as a short pulse of energy that seem to defy reason by persisting intact for an inordinate amount of time. The first recorded mention of solitons in modern scientific literature was that by J.S. Russell, who in 1834 noticed a bow wake of a canal boat propagate nearly two miles after the boat stopped, and began an intensive study of the phenomenon as a result. Soliton wave pulses travel in such a way as to almost completely cancel out the dominant "dispersion" effects most waves succumb to, and as a result suffer very little energy dissipation as they propagate. This has been modeled using several non-linear equations, including:

$$k_2 \; [\, d^2A / dt^2 \,] \; + \; [\, k_o A^2 \; - \; k_1 \; dA/dt \,] \;\; = \;\; 0$$

where "A" is the pulse amplitude and the "k's" are various term coefficients (i.e. constants). With a little reflection, we can see a very strong resemblance between this equation and the Schrödinger Equation:

$$-\hbar^2 / (2m) \;\; d^2 \psi_{(x,t)} / dx^2 \; + \; U \, \psi_{(x,t)} \;\; = \;\; i \hbar \;\; d \, \psi_{(x,t)} \, / \, dt$$

Bringing the right hand side over to the left, this becomes:

$$-\hbar^2 / (2m) \;\; [\, d^2 \psi_{(x,t)} / dx^2 \,] \; + \; [\, U \, \psi_{(x,t)} \; - i \hbar \;\; d \, \psi_{(x,t)} \, / \, dt \,] \;\; = \;\; 0$$

(replace the state function "ψ" with the amplitude "A" as well as the constants, and the two equations become nearly identical).

The strong similarity between these two equations implies that a traveling electron can be thought of as a soliton type object. In fact, that is exactly how we have been treating it: – as a superposition of a collection of energy waves, or in other words, as a "*wave-packet*", traveling through space with zero loss everywhere except while tunneling through a barrier.

A fair amount of research has been dedicated to the study of solitons, particularly in their application to fiber optics. The use of such low-loss waves in fiber optics could offer a way to extend the path distance considerably in a fiber link, something that would prove very useful as we grow our long-haul fiber optic networks around the world (see Chapter 8 for additional discussion).

<u>The Raman Effect</u>:

Our final topic in this chapter is one that often comes up in spectroscopy and fiber optics, known as the Raman Effect (named after Chandresenkhara Raman who first observed it in 1928, five years after it was predicted by Adolf Smeckal). The

Raman scattering effect is based on a relatively simple concept related to Quantum energy levels, but like many of the topics in this chapter, it cannot be explained using Classical physics.

The equations that describe how electromagnetic waves propagate (i.e. Maxwell's Equations), are *linear* equations (as opposed to, say, quadratic "x^2", or cubic "x^3", etc.). The fact that they are linear tells us that Electromagnetic energy does not "mix" or combine when it propagates through a "normal" (i.e. passive) medium, such as the vacuum of free space.

When electromagnetic energy propagates through a region containing a reasonably high concentration of atoms or molecules (e.g. a gas cell, a block of plastic, a long glass fiber, etc.), the atoms/molecules in the material affect the electromagnetic energy as it propagates. We mentioned in Chapter 2 that this affect is accounted for by the "permittivity" and "permeability" of the media (with each being a function of the material, and the frequency of the electromagnetic energy). Having introduced the concepts of the Quantum model, we are now in a better position to understand how atoms/molecules have this effect on electromagnetic energy.

From our previous discussion on resonance and absorption, we know that atoms and molecules effectively form miniature resonators, and when the energy of the electromagnetic wave happens to correspond to a resonant frequency of an atom or molecule in the media (i.e. the E&M photon's energy $E = hf$, equals the energy difference between two atomic or molecular energy states), that photon is absorbed. What then does the atom or molecule do with that energy? From Quantum studies, we know that there are *several* possibilities.

The most probable (i.e. the most common) response is that the atom or molecule almost immediately de-excites (within ~10^{-15} seconds). In the process, it re-radiates this energy by releasing a photon at the same frequency (but not necessarily in the same direction). This process is essentially a passive response and is referred to as *Rayleigh Scattering* (or in classical parlance, as "elastic scattering", since no energy is lost). This scattering mechanism is responsible for the sky being blue, and why sunsets are mostly red/orange, since light at sunset must pass through ~twice as much atmosphere, it scatters ~twice as much blue light in random directions in the process (leaving mostly only red light to reach the observer).

The next most likely possibility is known as *Raman Scattering* and occurs when the atom or molecule de-excites by going to some *other* level other than the level from which it started. In reality there are two distinct types of Raman emissions, known as *Stokes* and *Anti-Stokes* emissions.

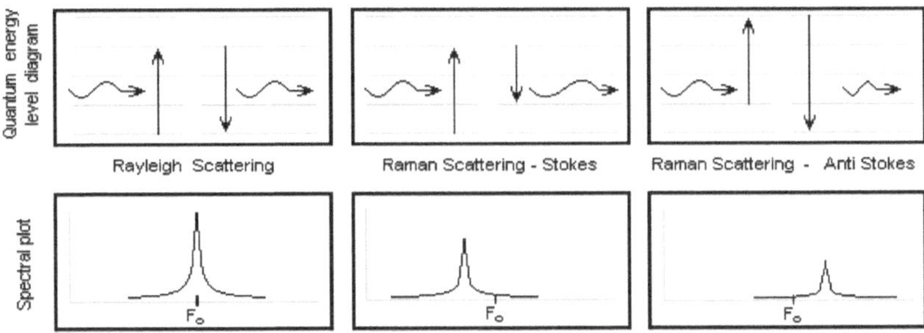

Figure 4.6 Rayleigh and Raman Scattering, including Stokes and Anti-stokes emissions.

If the atom/molecule de-excites to a valid level that is between the excited state and the starting level, it releases a photon that is at a *lower* frequency than the injected photon (in classical parlance, this would be considered an "in-elastic scattering" process, since the released photon has less energy than the original photon). This type of Raman scattering produces a *Stokes* spectral line (lower in energy/frequency than the Rayleigh line), and is the most probable type of Raman emission (since most atoms or molecules found in nature tend to be in the lowest, most stable energy state possible, as predicted by the Law of Entropy).

If on the other hand, the atom or molecule was already in an *excited* state for some reason *prior* to the encounter with our photon, the intruding photon could then further elevate that atom or molecule to an even higher state. From this higher level it may then de-excite to any number of levels below it. If it falls farther than it was elevated by our photon, it will release a photon of higher frequency than our original photon. This higher energy photon accounts for what is known as the *Anti-Stokes* spectral line.

From this description, it is clear that the Raman scattering process is not a "passive" or "linear" process. It cannot be understood using classical descriptions, such as the Lorentz "ball and spring" model or by using the standard Maxwell's Equations alone. It can only be explained using the insights the Quantum model provides.

By careful choice of gain media atoms or molecules, we can design a system to exploit the Raman Effect in order to create a stimulated lasing "gain" process that can be pumped by one wavelength laser, and which will then be de-excited by a lower wavelength photon when stimulated. We will discuss just such a system in Chapter 7 as it applies to a fiber optic link.

Chapter 4 Review questions:

1. What is the "Principle Quantum number", "n"?
2. What is the "Orbital Angular Momentum Quantum number", "l"?
3. What are "Sub-orbitals", and what are the restrictions on them?
4. What is the "Spin Quantum number", "m_s" ?
5. Why were these other sub-orbital/spin Quantum numbers introduced (why were they needed), and to what part of the electron energy do they correspond?
6. How does the electron occupation into these orbital states affect the chemical properties of each atom?
7. How does an electron being moved to a larger orbit store energy in the atom? Where does that energy come from?
8. What is the "Paulie Exclusion Principle", and how does it affect the chemical properties of each atom?
9. What is a Nobel Gas and why are they different from other elements?
10. What is the Dipole Selection Rule, and how does it help explain Metastable states?

11. How does Schrödinger's Equation help us describe a Quantum system?
12. What is an Eigenstate? An Eigenoperator? An Eigenvalue?
13. What is a "Stationary State", and what mathematical step did we use to introduce it?
14. How does an electron's energy change when it is near a potential?
15. What is "Tunneling" and what does classical physics tell us about it?
16. What did Einstein mean by saying "God does not play dice with the universe"?
17. Why do we use "Statistical Probabilities" to describe the outcome of a Quantum system?
18. What are solitons, and how are they analogous to our description of an electron as a "wave packet"?
19. Explain the Raman Effect.
20. What are "Stokes" and "Anti-Stokes" emissions, and what are their relationships to Rayleigh "scattered" emissions?
21. What is the mechanism responsible for the Raman "Stokes" and "Anti-Stokes" emissions?
22. How do we explain Raman Scattering using classical physics?
23. What is the probability of rolling four dice and all coming up the same (e.g. all ones)?
24. If an electron (mass = $9.11\text{x}10^{-31}$ kg) is moving at 100 m/s, what is its momentum (p=mv), and its de Broglie wavelength (λ)?
25. Calculate the kinetic energy of the electron in the previous question.
26. If that same electron is moving through a 100 volt potential, what is its total energy (KE + U)? (See footnote #11.)
27. If we have a free electron with the following state function: $\Psi_{(r,t)} = 1/\sqrt{(4\pi)}\ \delta_{(r,o)}\ e^{-i\omega t}$, what is the probability of finding it if we search the entire universe over all time? (Hint: the "Kronecker delta" implies the state function is zero everywhere except at r=0, where it equals one; the probability volume integral therefore collapses to a single point at r=0, making this one of the easiest integrations you may ever do.)

Chapter 5 Applied Quantum Physics: Atomic and Molecular Lasers

Now that we have developed some of the more essential concepts of Quantum Physics, we are now in a position to use them to help us more fully understand how a laser works. Since everyone loves hard, tangible examples, we will begin this chapter by discussing an example of a laser which uses single isolated atoms as the gain media, the Neon laser. We will then "improve" upon this basic design using what we have learned thus far to create the HeNe (Helium-Neon) laser. Along with the basic components of this improved design, we will also include a few simplified calculations to demonstrate how a fraction of the original energy in the electron discharge pumping mechanism is routed into a productive metastable state (with the remainder of the energy being distributed into a number of various loss mechanisms).

Following this HeNe example, we will transition into a discussion about molecular lasers, finding that combining atoms together effectively splits the electron energy levels. We then increase the number of atoms in our molecule considerably, splitting these energy levels still further, and in the process find that these split energy levels then form into energy "*bands*" within a solid, such as a semiconductor crystal lattice.

Using this description of energy bands in a solid, we will then introduce the semiconductor laser diode. We spend a little time describing (in basic terms) some of the fundamental aspects of the operation of diode lasers, highlighting a few important operational issues along with a brief mention of some innovative solutions, including the relatively new "Vertical Cavity Surface Emitting Semiconductor Lasers" (VCSELs).

HeNe laser [18]:

Though the Helium-Neon laser has two prominent elements in its name, only one is the lasing species of interest, namely the Ne atom (the same atom that is responsible for the characteristic orange glow of your favorite Neon night light). The Helium atom is added to the Neon gas as an auxiliary "impurity" with the specific purpose of helping to improve the pumping efficiency. The pumping itself is achieved through electron discharge, using a high voltage supply (~1–2 kV) applied between an anode and a cathode within the cavity containing the HeNe gas (1 part He to 10 parts Ne, at $\sim 10^{-2}$ atmospheres of pressure, i.e. ~100 N/m or 0.15 lb/in^2).

If we start with only Neon gas in our laser cavity, we find that some small percentage of the electrons in the discharge "arc" collide with the Neon atoms in the gas, transferring a small portion of the "arc's" kinetic energy into the atoms. Of this energy, only a very small fraction coincides with the exact energy required to elevate the Ne atom into the $3s_2$ level, which is the desired metastable state needed to lase. The rest of the transferred energy ends up in other non-productive areas.

In order to improve the efficiency of this simple design, we need a way to channel more of the energy from the electron discharge into the productive Ne $3s_2$ state. The simplest way of doing this is to add other types of atoms to the mixture which somehow help transfer energy into this state. This energy transfer between two atoms can best be understood by picturing two colliding billiard balls as an analogy. If we strike the cue ball and send it towards a second stationary ball, the fact that the cue ball is moving implies we imparted some amount of kinetic

---

18 Standard spectroscopic notation for energy levels typically uses the following format: $N\ ^{2S+1}L_J$, where "N" is the Bohr shell number, "L" is the orbital angular momentum, "S" is the spin angular momentum and "J" is the total combined angular momentum (J = L + S), and the "L" letter designation is: 0='s', 1='p', 2='d', 3='f', 4='g', etc. For example, the ground state for Helium is 1s, or 1^1S_o, while the first excited state is: 2s, or 2^1S_o, etc. Note that the following molecular Selection Rules apply: $\Delta L = \pm 1$ and $\Delta J = 0$ or ± 1.

energy into it (KE = ½ mv^2), while the stationary ball (i.e. v=0) has none. From experience, we know that as the cue ball hits the stationary ball, the cue ball slows down slightly as the second ball begins to move. The fact that the second ball is now moving indicates it has now gained some kinetic energy from this collision, while the cue ball lost some energy, which it transferred into the second ball upon impact.

If we knew the mass of each ball, exactly how much energy the cue ball had before impact (i.e. its speed), the elasticity of these balls, and the angle of impact into the second ball, we could use Newton's physics to predict how much energy would be transferred and at what angle both balls would then travel as they move away from the collision point[19].

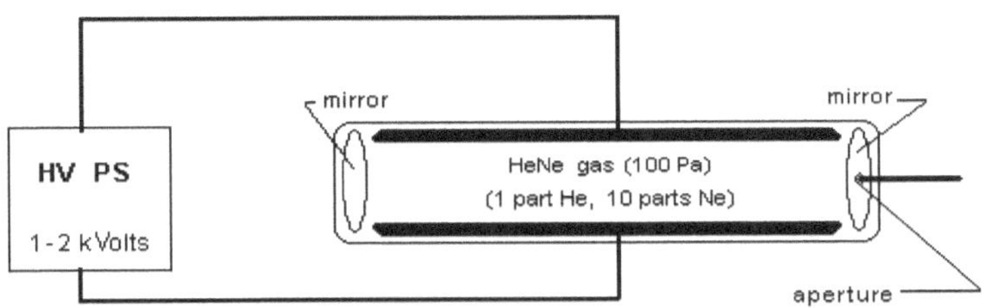

Figure 5.1 HeNe laser block diagram

In a simple approximation, we can say that a similar event occurs when two Quantum objects collide. The only significant difference is that in the Quantum world, the only way energy will be transferred is if it *exactly* equals one of the

---

19 This is the level of determinism Einstein expected from any solution physics provided to the Quantum world. The problem with this expectation, is that it was artificial, in that such textbook problems were simplistic. If we add one or two more balls to the mix at the collision point (e.g. a four-body collision problem), the Classical deterministic approach breaks down completely and is no longer able to develop a solution. If only a four-ball collision is impossible to solve, a "trillion-body" problem in a thin mixture of gases is infinitely worse.

discrete energy levels of the recipient atoms. This suggests that we could improve the efficiency of our Neon laser if we can find an "impurity" atom that has an energy level equal to the metastable energy state of the Neon atom.

As it turns out, Helium is just such an atom. The $2s_1$ state of the Helium atom holds roughly the same amount of energy as the $3s_2$ Neon state, and as a result, when these two atoms collide, the excited Helium atoms transfer energy into the Neon $3s_2$ state. Even with this improvement, the output power levels from the HeNe laser are typically not much more than a few milliWatts, but the 632.8 nm emissions are coherent and "relatively" monochromatic (we will discuss some effects that occur in the gain media that can "broaden" the frequency of the photons being emitted from a laser shortly).

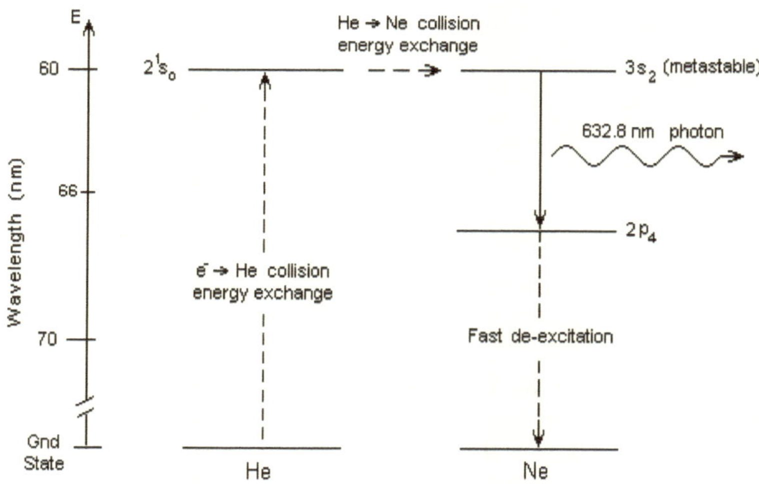

Figure 5.2 HeNe laser Energy level diagram.

Rate Equations for the HeNe Gain Media:

Since the Law of Conservation of Energy tells us that energy is always conserved, we know that whatever amount of energy we inject into the laser cavity (via electron discharge, or flash lamp pumping, etc.) has to all go somewhere. With that in mind, we can write a simple "balance sheet" type equation that will show the percentage of energy being directed into the productive metastable state, and the percentage of energy that is lost into other non-productive states (typically a large portion of our energy ends up lost into other "useless" energy states, or as cavity heating, etc.). These equations will allow us to describe the rate at which energy transfers into the desired atomic energy states, and are therefore known as "*Rate Equations*". To develop these equations, we will simply combine all of the pieces that help move our gain media towards the desired productive state, along with all the non-productive mechanisms that detract from that goal.

Once we have constructed our rate equations, we will be able to use them to express a special case known as the "*Steady-State*" condition, which occurs when we are injecting just enough raw energy into the gain media via the pumping mechanism to offset all the losses in the laser (i.e. energy diverted into non-productive paths) and thereby establish a continuous steady-state output.

We stress that the amount of energy being fed into the gain media at Steady State is more than just the amount of energy exiting the system via the laser beam. As mentioned during our opening remarks on Carnot engines, there are many parts to our laser "engine" which do not contribute to the production of photons at our desired wavelength. This is completely analogous to what happens in any other Carnot engine, including in our favorite family car (e.g. energy lost in friction, in generating a spark, in opening and closing exhaust valves, etc.). Some of the non-productive loss mechanisms in the laser include the excitation of atoms into a variety of other non-productive states that do not contribute to the production of photons at the desired wavelength.

Of the small fraction of energy that does make it to the desired metastable state, we will find much of that energy often fails to end up in our output beam, and is instead absorbed by impurities, or lost as heat in various cavity structures, etc. In fact, in the vast majority of cases, the overall process of pumping the gain media is a relatively inefficient one, requiring us to throw a great deal of raw energy at the system just to get some small portion of that energy to transfer into a productive metastable state and then into our output beam. Though there are some clever tricks we can play to help improve the transfer of energy into a desired state (and hence improve the overall efficiency of the laser), this whole process is effectively a statistical game of chance; the odds of hitting the desired state typically aren't that high (e.g. "1 in 50"). We therefore must compensate for those low odds simply by injecting as much energy into the system as we need to overcome the losses and achieve the desired output.

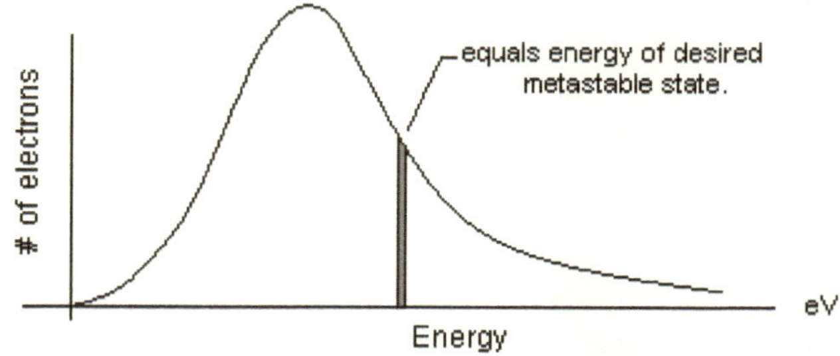

Figure 5.3 A plot of the distribution of energy in the electron discharge. In this plot we see only a very narrow fraction of electrons are at the right energy to match the excitation levels related to the productive metastable state.

These rate equations allow us to estimate not only the fraction of energy that successfully transfers into the desired atomic energy states, but they also allow us to specify the minimum pumping level or threshold ($p_{threshold}$) that we must achieve in order for the laser to reach a sustained steady state output. If the raw power we are injecting into this Carnot engine fails to operate above this minimum threshold level, the energy stored in the gain media will be rapidly depleted, and as a result the lasing process within the cavity will cease (this will also be a factor when we discuss multiple modes operating within a laser cavity, and where those modes which are $< p_{threshold}$ die out).

Let's begin by constructing a set of simplified rate equations for our Helium-Neon gas laser. To do this, we will need to define the following terms corresponding to our HeNe energy level diagram:

$N_{He\text{-}Gnd}$ = the number of Helium atoms in the ground state per cm^3;

N_{He2s} = the number of Helium atoms in the excited "2s" state per cm^3;

dN_{He2s} / dt = the rate of change of He_{2s} (into Ne_{3s} and/or back to ground):

a = the pumping rate (collision energy transfer coefficient between the excited electrons in the discharge arc elevating He atoms to the He_{2s} state);

b = the collision energy transfer coefficient between He_{2s} and Ne_{3s};

c = the rate of non-productive de-excitations of He_{2s} atoms;

Focusing on these main factors of interest in the normal pumping life-cycle, we have:

$$dN_{He2s} / dt = a\,N_{He\text{-}Gnd} - b\,N_{He2s} - c\,N_{He2s}$$

In words, this equation tells us the change in the number of excited He atoms depends on the number of He atoms in the ground state, times the efficiency of the pumping process (how well it excites ground state He atoms into the desired excited state He_{2s}), *minus* those atoms losing energy (transferring it to the other non-productive states).

Similarly, we can write an equation for the change in the number of Ne atoms in the excited 3s state:

$$dN_{Ne3s} / dt = b\,N_{He2s} - f\,N_{Ne3s} - g\,N_{Ne3s}$$

where "f" = the rate of energy leaving the Ne_{3s} state through non-productive de-excitations (e.g. spontaneous emissions, collisions with the cavity wall, etc.), and "g" = the rate of energy leaving the Ne_{3s} state through productive Stimulated emissions.

We note that since these two processes are interwoven, they form a set of *coupled* equations, which can typically be a little challenging to solve. In the steady state condition however, there is *no change* in the *number of atoms in the excited state* and therefore both dN/dt's = 0. as a result, when steady state is achieved, the second equation above becomes:

$$0 = b\,N_{He2s} - f\,N_{Ne3s} - g\,N_{Ne3s}$$

which can be re-written as:

$$b\,N_{He2s} = f\,N_{Ne3s} + g\,N_{Ne3s}$$

plugging this back into the first equation:

$$0 = a\,N_{He\text{-}Gnd} - f\,N_{Ne3s} - g\,N_{Ne3s} - c\,N_{He2s}$$

or:

$$a\,N_{He\text{-}Gnd} \geq f\,N_{Ne3s} + g\,N_{Ne3s} + c\,N_{He2s}$$

In other words, to maintain a steady state output, the pumping process ($a\,N_{He\text{-}Gnd}$) must be at least equal to the amount of energy leaving the cavity in the form of photons, *plus* the amount of energy needed to offset the losses due to energy being transferred somewhere other than the He2s state (reflected by the fact that "a" is not 100%), being lost out of the He2s state into something other than the Ne3s state, and finally energy lost out of the Ne3s state (into something other than a productive photon).

Simple Conservation of Energy tells us that the amount of energy entering must equal the amount of energy being ejected + that being lost. If it helps, think of the whole process as a balance sheet for your typical home or business: so much going into the budget, and so much going out (where the amount going out = useful expenditures + waste).

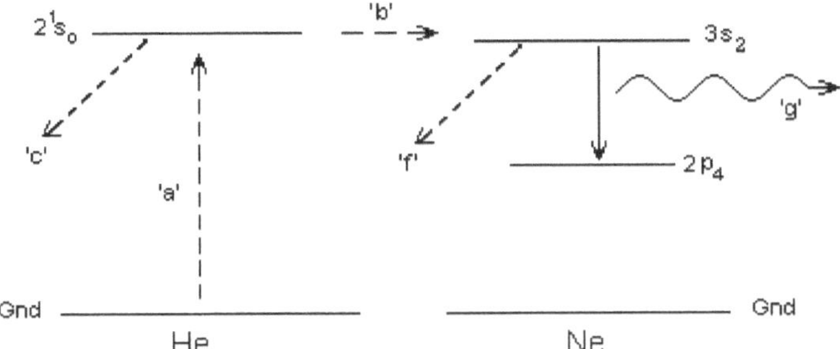

Figure 5.4 HeNe Energy diagram showing "balance sheet" coefficients used in the rate equations.

Example:

Given: "b" = 20%, "c" = 65%, "f " = 60%, "g" = 30%, $N_{He\text{-}Gnd} = 10^{18}$ (atoms/cm^3) and $N_{Ne3s} = 10^{15}$ (atoms/cm^3), find: N_{He2s} and the pumping coefficient "a" required to maintain this steady state level. From this information, estimate the amount of discharge current (electrons per second) needed to maintain a steady state output from the laser.

$$N_{He2s} = (f\,N_{Ne3s} + g\,N_{Ne3s})\,/\,b$$

$$N_{He2s} = (60\% + 30\%) \times 10^{15}\ (\text{atoms/cm}^3)\,/\,20\%$$
$$= 4.5 \times 10^{15}\ (\text{atoms/cm}^3)$$

Having found N_{He2s}, we can now solve for "a":

$$a = (f\,N_{Ne3s} + g\,N_{Ne3s} + c\,N_{He2s})\,/\,N_{He\text{-}Gnd}$$

$$a = [\,(60\%+30\%) \times 10^{15} + 65\% \times 4.5 \times 10^{15}\,]\,/\,10^{18}$$

$$= 0.4\ \%$$

Which is the minimum pumping rate to elevate the He atom from the ground state into the He_{2s} state to sustain steady state operation. The next question to be addressed is how great of an electron discharge do we need in order to achieve this steady state condition? Since our transfer coefficients are intended to account for all losses (albeit rather crudely), we will assume for now that all energy injected into the system by electron discharge is conserved. Therefore we can find the required number of electrons needed (N_{input}) with the following:

$$(N_{input}) \text{ (xfr efficiency) } = N_{out} = g\, N_{Ne3s}$$

or:

$$N_{input} = (g\, N_{Ne3s}) \,/\, (a * [1\text{-}c] * b)$$

$$= (30\% \times 10^{15}) \,/\, (0.4\% \times [1\text{-}65\%\,] \times 20\%)$$

$$= 1.1 \times 10^{18} \text{ electrons / sec}$$

$$= (1.1 \times 10^{18} \text{ e's/sec})/(0.62 \times 10^{19} \text{ e's / Coulomb})$$

$$= 177 \text{ mAmps minimum discharge current}$$

(where we used the abbreviation "e's" for "electrons".)

In terms of Watts, we know that the energy from a single photon is: E = hf, and therefore, the energy from "n" photons is: E = nhf. If we ideally assume *all* 10^{15} N_{Ne3s} atoms provided photons to the output, and assume a laser cavity volume of 10 cm^3, in a one second slice of time these photons can be equated to Watts (W = J/s):

$$E = n\, h\, f \times \text{volume}$$

$$= (10^{15} \text{ p's/cm}^3)\ h\ (300 \times 10^6 \text{ m/s} / 632.8\text{nm})\ (10 \text{ cm}^3)$$

$$= 3.11 \times 10^{-3} \text{ Joules released in 1 second}$$

$$= 3.11 \text{ mW}$$

(where we used the abbreviation "p's" for "photons".)

Additionally, we should also include some account of the losses in the electronics driving our discharge (i.e. for transformer losses, heat lost in resistances, etc.). For the sake of our arbitrary example, we will assume a 65% efficiency for our imaginary power supply. In that case, we would need to inject *at least* 273 mAmps (= 177mA/65%) into the power supply (or roughly 273mA x 110VAC = 30 Watts) to get 3.1 mW out:

Efficiency = Pout / Pin = 3.1mW /30W = 0.01% !

Granted, we made a number of simplifying assumptions in these calculations, but this analysis does provide a rough idea of the typical operational efficiency of a HeNe laser. To be brutally honest, HeNe lasers are not very efficient, due primarily to the limited mechanisms available that contribute to the desired metastable state.

If we wish to develop a more efficient laser (i.e. raise the output power while running at a reasonable cost), we will need to find a gain media that is a little more effective in its refinement of the raw input energy towards achieving a strong coherent output beam. To that end, we will now step up our discussions a notch to consider the behavior of the typical materials used in the fundamental lasing mechanism.

<u>Molecules as Gain Media</u>:

So far we have only talked specifically about electrons around atoms, and only vaguely eluded to the use of Quantum Physics in connection to molecules. Obviously, since electrons and atoms form the constituent building blocks of molecules, all the Quantum machinery that apply to atoms, by default apply to molecules as well. In addition to the lasers designed around the metastable states of isolated atoms (e.g. HeNe), a vast number of popular laser types in use today exploit the metastable states of various species of molecules, including the high power CO_2 laser. To explore these systems, we will now take what we know about atomic systems and apply it to construct a working description of a molecule.

To begin our discussion on molecular modes, we point out that anytime two atoms are brought together to form a molecular bond, they do so out of a need to complement each other's outer shell and thereby form a more stable configuration. For example, when two Hydrogen atoms combine to form a

Hydrogen gas molecule (H_2) they effectively share their electrons between them to complete each other's "1s" sub-orbitals. The resultant molecular bond is therefore referred to as "σ" bond (Greek letters are typically used to designate the molecular equivalent of atomic sub-orbitals, with "σ" being the equivalent Greek letter "s"). The next higher molecular bond is the "π" bond (the Greek equivalent to "p"), etc.

Since the Paulie Exclusion Principle forbids two electrons from occupying the same state (i.e. energy level), the two electrons in each sub-orbital take alternate "spin" states m_s, one being in a "spin up" state, and the other in a "spin down" state. As a consequence of the Exclusion Principle, these "split" molecular states have slightly different energy values (as evidenced by spectral data). We also note that though the initial "sharing" of the first few electrons does create a more stable molecule (which we show as a bond that is lower in energy than either of the shells of the isolated atoms; see Figure 5.5), the accumulation of additional electrons can induce repulsion between the two atoms (due again to the Paulie Exclusion Principle), creating something of an "anti-bond". If the attractive sharing is stronger than the repulsion, a stable molecular bond is formed. If on the other hand, the anti-bond cancels, or is greater than the attraction, the bond is unlikely to form (or if it does form, soon disassociates).

To estimate the likelihood that such a bond will form, we use the following formula:

$$b \quad = \quad \frac{1}{2}(n - n^*)$$

where "n" is defined here to be the number of electrons in the more stable bond, and "n*" is defined to be the number of electrons in the "anti-bond". A positive bond strength "b" indicates the bond is likely, while a zero or negative "b" indicates it is unlikely. From the diagram in Figure 5.5, we see that diatomic Hydrogen gas (H_2) is a more stable configuration than isolated Hydrogen atoms alone (H), indicating the formation of diatomic Hydrogen is favored in nature. Diatomic Helium gas (He_2) on the other hand is not a more stable configuration, and therefore is *not* likely to be found in nature. Similarly, we see that diatomic Oxygen gas molecule (O_2) is a more stable configuration for Oxygen, as opposed to isolated Oxygen atoms (O).

Exercise: Using the previous bond strength equation, determine if He_2^+ (i.e. Helium ion with only 3 electrons) is likely?

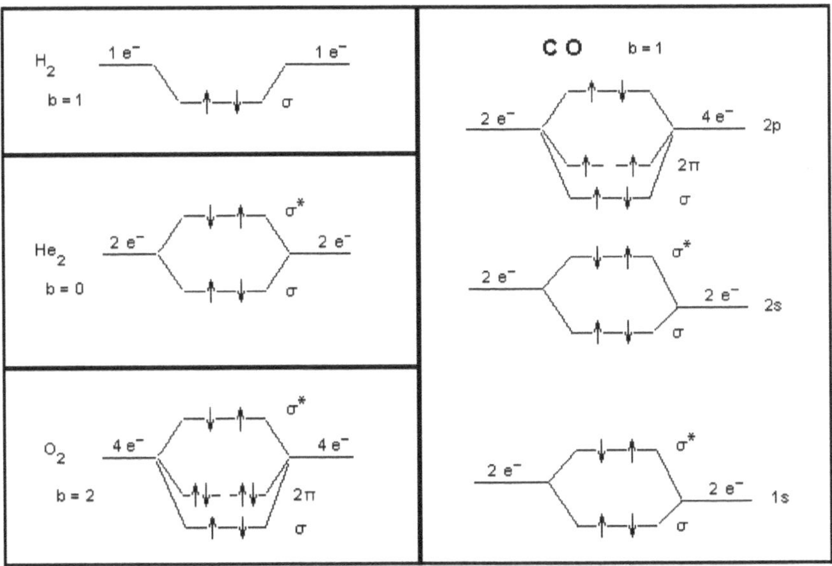

Figure 5.5 Examples of molecular bond formation diagrams for several various gases.

When atoms do coalesce into molecules, they form what can be very elaborate and extended structures, each with its own characteristic shapes, bond strengths, and hence spectral signatures. Due to the shared electron bonding between atoms, there is a definite *preferred arrangement* in their extended molecular structures. This coupled with the "elastic"/"spring-like" force of the bond implies a "restoring force" exists in the molecule very similar to our "ball and spring" Simple Harmonic Oscillator (SHO) model. As a result, anytime the atoms in a molecule are disturbed (e.g. by a collision with other molecules, or a stream of energetic electrons, etc.), the restoring force causes the disturbed molecule to quickly "spring back" towards its preferred shape. However, as with any free-form object with a spring action, the restoring force causes the disturbed atoms to overshoot their preferred "resting" position, causing them to begin oscillating back and forth around this preferred shape. This SHO behavior can be modeled using something similar to the Lorentz "ball and spring" model, by viewing the atoms in the molecule as if tethered to each other by a spring.

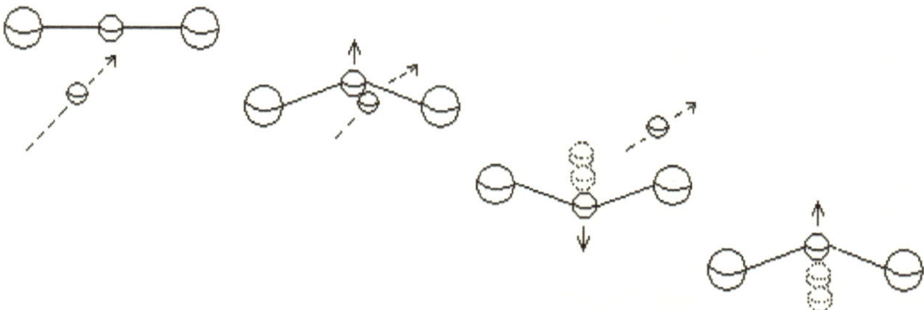

Figure 5.6 One possible vibration excitation of a CO_2 molecule.

As we saw in our previous example of the CO_2 molecule (see Figure 2.2 and Figure 5.6), the disturbing force of any "impact" event injects energy into the molecular spring system, causing it to vibrate/oscillate in a variety of different modes, including bending, stretching, compressing or rotating (typically some combination of these).

For a molecule to radiate, it must have a *non-symmetrical* charge distribution. To understand why, recall our discussion in Chapter 2 related to the "electric dipole" in a SHO. Any time a molecule's charge is not symmetrically distributed (for example, the electron distribution around a bent water molecule is clearly not symmetrical), the molecule can be thought of as having a polar charge distribution with the net positive and negative charges at "opposite" sections of the molecule. This separation of net charges across the molecule effectively creates a *dipole* moment "p", which is a function of the net charge "q" and the distance separating the net charges "**r**":

$$\mathbf{p} = q\,\mathbf{r} \qquad \text{electric dipole}$$

From basic Electrodynamics, we know that anytime a charge distribution moves, it radiates a field (see Chapter 6). Just how many radiating modes a molecule has, therefore depends on how many ways its charge distribution can vary. Consequently composition, shape, and distribution of charge determine whether or not a molecule has a vibrational or rotational spectra.

With a little practice, one can easily determine which molecules will and which will not have either a *rotational* or a *vibrational* spectra, based on the *symmetry* of the molecule (and therefore the symmetry of its charges). For example, H_2 ("H–H") has a symmetrical charge distribution which remains symmetrical even during vibration and rotation. H_2 therefore does not exhibit either types of spectra. CO on the other hand is polar, and thus has both a vibrational and a rotational spectra.

To estimate the number of *vibrational* modes a molecule might have, we can use a simplified rule of thumb, which is based on the number of atoms in the molecule ("n") and the symmetry of their arrangement. Each atom in the molecule has three degrees of freedom, giving us a maximum of "3n". However the bonds between its atoms effectively restrict their freedom of motion and thus reduce the degrees of freedom. Therefore, considering the symmetry of the molecule (i.e. how its charge distribution changes under *translations,* and under *rotations*), we find we must subtract one degree of freedom for *translations* along each axis.

In addition, if the molecule is *non-linear* (e.g. H_2O), we must also subtract one degree of freedom for each possible rotation around each axis. If the molecule is *linear* (e.g. CO_2), a rotation around its long axis produces no change, and therefore we do not subtract a degree of freedom for that axis. Therefore the total number of degrees of freedom are:

$$N = 3n - 6 \qquad \text{non-linear molecules}$$

$$N = 3n - 5 \qquad \text{linear molecules}$$

For example, we find H_2O has *three* vibrational modes, while the linear CO_2 molecule on the other hand has *four*.

In order to induce either a *vibration* or *rotation* into a molecule, energy must be injected into it from an external source. A *collision event* with another energetic molecule will transfer some kinetic energy from one into the other (much the same way a cue ball imparts energy into a stationary billiard ball on impact). The molecule can also absorb energy from an external *Electromagnetic field,* provided of course that the injected energy corresponds to one of the discrete energy modes of that molecule. However from our previous comment on Electromagnetic fields, we note that before a molecule can radiate, its net charge distribution must *change*. If the molecule has a non-uniform charge distribution (i.e. a "polar" molecule such as HCl), that requirement is easily met anytime it moves. The net charge distribution around non-polar molecules such as H_2, and N_2, (etc.) does not change as they oscillate, and consequently they do not exhibit vibrational or rotational spectra.

In the case of *rotational* modes, we note that the rotational motion in the presence of an E&M filed is due to the Lorentz force (see Chapter 2) the field exerts on the net dipole of the molecule, causing it to rotate to align with the external field. As its charges rotate, they radiate. Another way of looking at this release of energy is to note that when the molecule rotates, its potential energy decreases as it aligns to the external field, implying some energy is extracted in the process. This is analogous to the loss in potential energy when a boulder rolls down a hill, or when an electron falls from a higher orbit towards a lower orbit around an atom, (etc.).

To underscore this point, recall our analogy from Chapter 3 in which we stored energy in water by pumping it up a hill against the force of gravity, which energy it then released as it flowed downhill (as it succumbed to that force). Moving the electric dipole of the molecule through the potential of the external electric field also represents work (analogous to moving mass through a gravitational field). In the case of a dipole in an Electric field, the energy change is:

Energy = **p • E**   (Dipole energy in an Electric field)

Again, we stress that not only must the molecule form a dipole, this dipole must also *rotate* (i.e. undergo a *change* in the orientation of the dipole) before any rotational energy is released in the form of a radiated photon. If on the other hand the molecule does *not* form a dipole (e.g. N_2 or O_2), it will *not* rotate in the presence of an external field, and hence it will not have a *rotational spectra*.

Bear in mind that these vibrational and rotational behaviors are *only part* of the possible energy storage modes we have discussed so far (the other being in the atomic orbits). Consequently whether the molecule stores energy in a *vibrational* or *rotational* mode or not, it will still present a higher frequency spectral signature via energy stored in its atomic orbits.

While on the subject, it is worth noting that the energy involved in each of these energy storage modes tends to be dramatically different. In other words, there is something of a "hierarchy" in the energy storage modes in atoms and molecules that relates directly to the difficulty in changing the atom's/molecule's charge distribution (i.e. the strength of the "elastic" restoring force involved). That hierarchy is:

1) electron-atom bonding (corresponding to visible and x-ray E&M waves),

2) molecular vibrations (corresponding to InfraRed light) and

3) molecular rotations (corresponding to radio waves in the GHz region).

Since causing a molecule to rotate or tumble in space requires very little energy, this third energy storage mechanism is the weakest of the three. As a result, it only requires energy in the GHz range of the ElectroMagnetic spectrum to cause the molecule to rotate or tumble (E=hf). The bonds that hold molecules together are a little stronger, requiring a bit more energy to cause a molecule to vibrate than it does to cause it to tumble. As a result, this second energy storage modes requires somewhat higher electromagnetic energy to activate, typically in the InfraRed range of the spectrum. The binding of the electron to the atom on the other hand, requires considerably more energy to disturb, particularly in the case of electrons in the lowest shell. As a result, these energy storage modes require the greatest amount of energy to change them, i.e. visible, Ultraviolet and in the case of electrons in the closest shell, x-rays photons. In other words, this "hierarchy of energies" implies:

$$E_{e-} > E_{vibration} > E_{rotation} \qquad \text{or}$$
$$F_{e-} > F_{vibration} > F_{rotation}$$

The total energy stored in a molecular system is:

$$E_{total} = E_{e-} + E_{vibration} + E_{rotation}$$

Up to this point in our discussions, we have described both the *atomic orbital* modes, and the molecular *rotational* modes of a typical atomic or molecular "system". There are however still a few things that we need to discuss about the molecular *vibrational* storage modes.

As with the rotational modes, a molecule can be driven into vibration by two means: 1) collision with another energetic Quantum object (e.g. an excited electron, atom or molecule), or 2) it can be forced into a vibration by the presence of an external E&M field with just the right wavelength, since vibrational modes (like all other Quantum mechanisms) have very *discrete* Quantum levels.

As it turns out, the "restoring force" of the molecular bond "spring" can be modeled reasonably well using an approach similar to the Lorentz "Ball-and-Spring" model introduced in Chapter 2, since the restoring force of a molecular bond behaves very much like a mechanical spring. To that end, we typically describe the strength of the molecular bond "spring" using a spring strength constant "k" as we would for a mechanical steel spring. To illustrate this approach, we show several molecular bond types in Table 5.1 with their approximate "k" values. We note however, that these approximate values are correct only for the bonds as simple, *isolated* molecules, and must be modified if transplanted into a more complex molecule due to the electrostatic influence of the other atoms/bonds in that molecule.

Table 5.1 Sample of simple molecular bonds.

bond	molecule	k (N/m)	ω (1/cm)
H–F	HF	965	4140
H–CL	HCl	515	2995
N=O	NO	1594	1900
C=O	CO	1902	2170

Figure 5.7 H_2O and CO_2 spring models.

Borrowing from our discussion in Chapter 2, we see that the oscillation frequency of a mass-spring system is directly related to the strength of the spring, and is inversely related to how massive the object is that the spring is trying to move:

$$\text{Freq}_{\text{(MHz)}} = \text{SQRT}(k/\mu)$$

where "μ" is the "reduced" mass of all the objects involved in the mass-spring system (defined below). However, since frequency in Megahertz is typically a very large number, most chemistry and spectroscopy data use a slightly different system of units based on centimeters rather than meters, which produces a "frequency" in "inverse centimeters", and which is referred to as the "wavenumber"[20]:

$$\omega_{(1/cm)} = SQRT(k/\mu) / (2 \pi c')$$

where $c' = 30 \times 10^9$ cm/sec, and "μ" is the combined total mass of the molecular system as a single "reduced mass" at the center of gravity of the combined system:

$$\mu = m_a m_b / ([m_a + m_b] N_{advogadro})$$

(with $N_{Advogadro}$ in grams equal to: 6.023×10^{23} atoms/mole.)

Example: Carbon Monoxide (CO) gas spectra:

Using Table 5.1 and the molecular weights of Carbon (12.0), and Oxygen (16.0), find the *fundamental* frequency for the single vibrational mode (in wavenumbers) and the wavelength (in meters) for Carbon Monoxide:

$$\mu = (12)(16) / ([12+16] N_{advogadro}) = 1.139 \times 10^{-26} \text{ kg}$$

$$\omega_{(1/cm)} = SQRT(1902 / \mu) / (2 \pi c') = 2170 \text{ cm}^{-1}$$

$$\lambda = 1 / (100 \times 2170 \text{ cm}^{-1}) = 4.6 \times 10^{-6} \text{ meters}$$

Example: The Nitrogen-Oxygen (NO) gas spectra:

Repeat the above calculations for NO:

$$\mu = (14)(16) / (14+16) N_{Advogadro} = 1.24 \times 10^{-26} \text{ kg}$$

$$\omega_{(1/cm)} = SQRT(1594 / \mu) / (2 \pi c') = 1903 \text{ cm}^{-1}$$

$$\lambda = 1 / (100 \times \text{ cm}^{-1}) = 5.25 \times 10^{-6} \text{ meters}$$

---

[20] The conversion of wavenumbers (in cm^{-1}) into wavelengths in meters is: $\lambda = 1 / (100 * wv\#)$.

In these examples we found the *fundamental* vibrational frequency for each molecule. However we note that there are multiple energy levels for each vibrational mode (analogous to the atom itself having multiple energy levels). it is therefore possible to excite these molecules into higher excitation levels above the fundamental. As with all Quantum systems, these molecular vibration energy levels are all *discrete*. Furthermore, we find that each level is separated by "$\hbar\omega$" according to the following approximation:

$$^{(\text{vib mode})}E_v = \hbar\omega\,(v + \tfrac{1}{2}) \qquad v = 0, 1, 2, \ldots$$

Zero Point energy and Degenerate modes:

Note that even at v=0 for each vibrational mode, the oscillator still has some minimum amount of energy (even at absolute zero), known as the "*Zero Point energy*", and is unique to the Quantum description of all such Quantum systems. This "Zero Point energy" in Quantum objects is effectively a result of the Heisenberg Uncertainty principle, which forbids Potential energy (a function of position) and Kinetic energy (a function of velocity) to both be zero simultaneously. As a result, all Quantum systems will always have some *minimal*, non-zero energy, even at absolute zero. Because the classical mindset does not include any consideration for the Heisenberg Uncertainty principle, the classical model does not show this minimal residual energy.

Recall that our previous estimate for the number of vibrational modes for CO_2 predicted *four* possible modes. However when we examine the spectra for CO_2, we can only detect *three* such modes. The reason for this is that the "y" bending mode and the "z" bending modes are identical except for their arbitrary alignment with our coordinate system axes. And since reality should not depend on our choice of reference frames, we find that they both have exactly the same amount of energy, and hence they radiate exactly the same spectral signature. Such energetically identical modes are known as

"*Degenerate*" modes, and in our spectral measurements are indistinguishable from each other.

Carbon Dioxide (CO_2) has 3 distinct (non-degenerate) vibrational modes (see Figure 2.2):

v00	Symmetrical stretch mode
0v0	Bending mode
00v	Asymmetric stretch mode

Therefore, the total vibrational energy stored in the molecule is simply the sum of the energies stored in all the individual modes:

$$^{Total}E_{vib} = {}^{(v00)}E_{vib} + {}^{(0v0)}E_{vib} + {}^{(00v)}E_{vib}$$

$$= \hbar\omega\,(v_{v00} + \tfrac{1}{2}) + \hbar\omega\,(v_{0v0} + \tfrac{1}{2}) + \hbar\omega\,(v_{00v} + \tfrac{1}{2})$$

$$= \hbar\omega\,(v_{v00} + v_{0v0} + v_{00v} + {}^{3}/_{2})$$

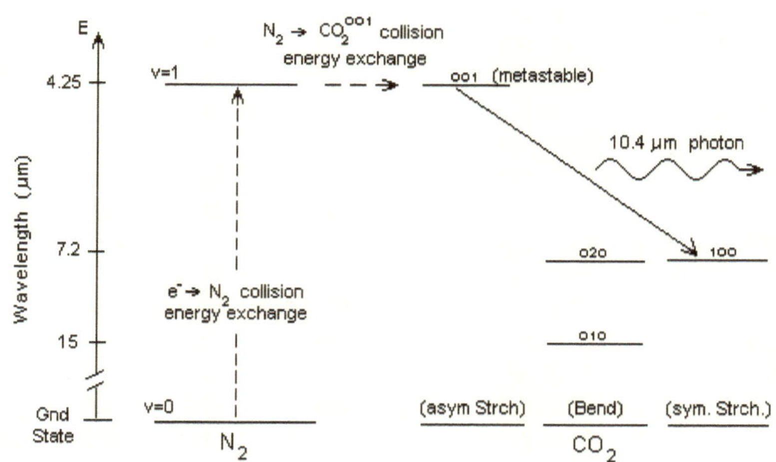

Figure 5.8 CO_2 molecular vibration state diagram.

Of these three vibrational modes in the Carbon Dioxide molecule, it is the "001" mode (the Asymmetric Stretch mode) in particular that interests us, since this mode is the metastable state which provides our lasing action for the CO_2 laser. The goal therefore, is to channel as much energy from our external pumping mechanism (an electron discharge) into this fundamental mode as possible.

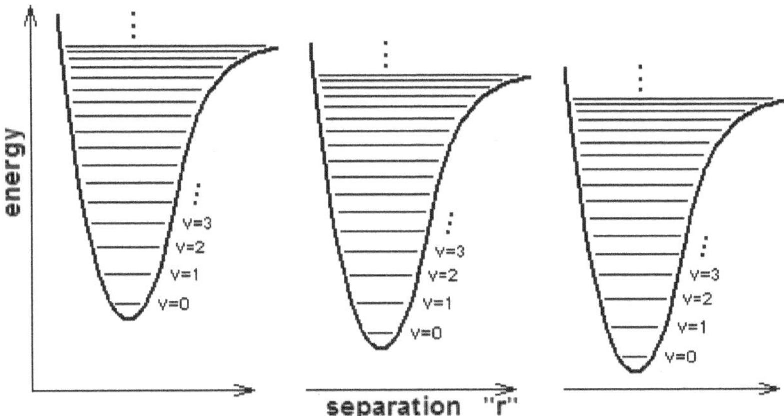

Figure 5.9 CO_2 vibrational mode Energy plot (not drawn to scale). Note the Zero Point Energy at v=0 for each mode lies slightly above the bottom of the parabola, i.e. even in the zeroth level, the molecule retains some amount of energy. Notice also that the parabolic shape and the separation between energy levels begins to distort with higher vibrational energy injected into each mode, asymptotically approaching disassociation.

As with the HeNe laser, we find that by using just the right "impurity" atoms in the CO_2 gas mixture[21], we can significantly increase the amount of energy channeled into the 001 mode, and thus increase the efficiency of the CO_2 laser. The most common impurity atom used in the CO_2 mixture, is

[21] A variety of gas mixture "recipes" exists for the Carbon Dioxide laser, depending on the application and preference of the manufacturer. As an example: CO_2: 8%, N_2: 10%, CO: 4%, H_2: 0.4% and He: 77.6% at approximately 0.01 atmospheres (10^3 Pascals) of pressure. The use of CO is optional, and serves to convert any freed O_2 back into CO_2, since O_2 degrades the pumping process as well as promotes corrosion of the discharge electrodes.

Nitrogen (N_2, which with two atoms, has 3(2) – 5 = 1 vibrational modes). As it turns out, Nitrogen's "v=1" vibrational mode just happens to resonate at about the same frequency as the CO_2 "001" mode, allowing it to transfer energy into this mode very easily upon collision with the Carbon Dioxide molecule (see Figure 5.8).

Drawing upon our discussion in Chapter 2 again, we know that the energy stored in a harmonic oscillator is proportional to the square of the distance the spring is disturbed from its normal resting position. This can be used to approximate the vibrational energy stored in molecules using the following equation:

$$PE_{spring} = \frac{1}{2} k x^2$$

In other words, the more we pull apart a spring (i.e. as "x" increases), the more potential energy we store in the spring, which it will demonstrate the moment we release it. We note that this equation describes a parabola. Therefore, when we plot all of these various energy levels in each vibrational mode, they take the form of a range of discrete levels within a parabola; and since there are *three* distinct vibrational modes for CO_2, we have three such parabolas, as shown in Figure 5.9.

Note that in reality, the shape of the curve is slightly distorted from the ideal parabola at the extreme range of energy, as we reach the edge of the validity of our approximated parabolic model. At extreme energy levels, the molecule is eventually torn apart when it is subjected to energy levels that are greater than the bond strength holding the atoms together. The net result of these effects is a flaring out of the parabola as the injected energy asymptotically approaches disassociation at these extreme ends.

Semiconductor lasers:

Thus far in our discussions, we have described matter in its simplest elemental forms, beginning with single isolated atoms and then branching out into the slightly more complex system of several atoms bonded together to form a simple molecule. These simple [isolated] configurations give us the luxury of describing the lasing effect under near perfect isolation in the form of a single atom or simple molecule in a rarefied gas. This implies very little influence exists between neighboring atoms (and their associated electrostatic fields), which would otherwise complicate things considerably.

Having developed most of the essential concepts and models related to lasing using these simpler configurations, we are now in a position to introduce the more complex system composed of a crystalline solid. In these more complex systems, the presence of neighbor atoms in such close proximity to each other results in splitting the bond energy levels far more than what we saw in simpler molecular bonds. As a result, a closed solution of Schrödinger's equation describing the material is no longer possible, and we now have to rely on the general concepts and models developed in these simplified cases to provide a more qualitative model to guide our understanding in the behavior of these more complex systems.

When we brought two isolated atoms together to form a simple diatomic molecule, we found that the mutual electrostatic interaction between these two atoms effectively split the individual atomic orbit into two distinct levels. We used the terms "bond" and "anti-bond" (i.e. "σ" and "σ*") to describe this "splitting" influence on the bonding action, noting that this splitting effect is a consequence of the Pauli Exclusion Principle which forbids any two electrons from occupying the same energy state.

If we add additional atoms to such molecules, we find that each time another atom is added to the structure, these individual energy states are split once more, again due to the Exclusion Principle. Consequently, anytime we combine "N" atoms together to form a solid, we find that each individual atomic energy state is split into "N" different, *distinct*, energy levels.

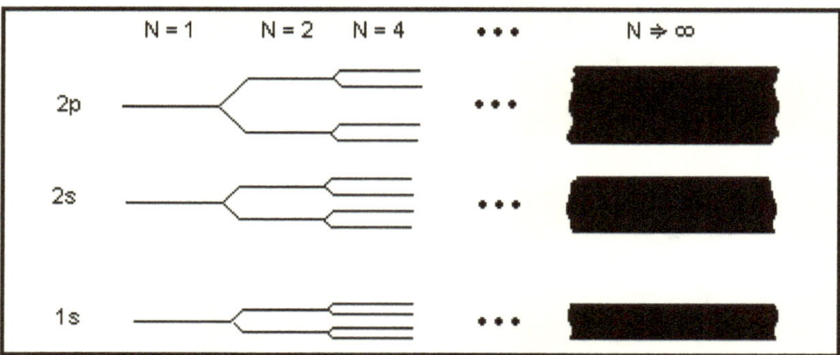

Figure 5.10 Atomic energy levels splitting into "bands" as the number of atoms combined together into a molecule approaches infinity (i.e. in a solid).

In the limit as "N" approaches a typically large number of atoms (> 10^{20}), we find that the splitting of individual energy states *broadens* the energy levels into an expanded range of "N" extremely fine and *closely spaced* levels. In effect, each of the very narrow and precise atomic energy levels discussed in the previous sections almost "merge" into a "smeared" range of possible energy values, referred to in the literature as an energy "*band*".

Since the basic Quantum rules governing each individual atomic component remain valid (e.g. the Paulie Exclusion Principle and the quantization of charge of each individual component, etc.), we infer that the regions between each enlarged shell or band remain "forbidden" regions, creating what is referred to as "Band Gaps" between each valid band, each with well known potential differences characteristic to each type of material.

With individual atoms, we found that the chemical behavior or interaction of the atom was heavily dependent on the occupation of the outer shell. This is also true in solids, since the outer most bands contribute the charge carriers (e.g. electrons) which define the solid's electrical properties. In addition to the number of electrons in the outer most band, we find that the *size* of the gap between the outer bands has a significant impact on the electrical properties of the solid as well.

The bands that interest us most are the "*Valence band*" (which is the highest filled band) and the "*Conduction band*" (which is the band above the Valence band, and is the band most responsible for the exchange of charge carriers, i.e. current flow throughout the solid). If the gap between the Valence band and the Conduction band is *narrow*, then it is much easier for electrons to be elevated up from the Valence band into the Conduction band, enabling the material to more easily supply charges to any overall current flow through the solid. Conversely, if the gap between these two outer bands is *large*, then it is much harder for the solid to contribute electrons to any current flow, making it more of an *insulator* than a *conductor*.

For example, the gap between the Valence and Conduction bands in pure diamond is 5.4 eV, while in pure silicon it is only 1.1 eV. This implies that it is *much* more difficult for electrons to migrate up to the Conduction band in pure diamond than it is for pure silicon; hence *pure* diamond (made of carbon) makes a great insulator, while *pure* silicon makes a fairly good insulator as well.

By comparison, the band gap between these two levels in a metal is virtually 0 eV, and hence electrons flow very easily across the boundary which separates the Valance band and the Conduction band. The close proximity of the Conduction band to the Valence band in metals makes it very easy for metals to contribute electrons to any current flow through the solid. As a result, metals make great conductors of electricity.

<u>The Fermi Level</u>:

Electrons belong in a class of atomic particles known as "Fermions". All Fermions obey the Paulie Exclusion Principle, which as we mentioned, forbids any two Fermions from occupying the same energy level. It is this principle that defines the energy levels around isolated atoms, as well as the energy bands in solids.

As electrons distribute themselves throughout the available energy bands in a solid out towards the top of the

Valence band, they form what can be thought of as a "*Fermi sea*". Before an electron can make it into the Conduction band and thus contribute to any current flow, it must first rise to the "surface" of this Fermi Sea. Therefore this "surface" energy level, known as the "*Fermi Level*", helps determine how well a material will conduct electricity. In other words, the location of the Fermi Level relative to the Conduction band affects the likelihood of electrons migrating from the Valence band into the Conduction band and thus contributing to the conduction current. At absolute zero, the Fermi Level is at the top of the Valence band. At warmer temperatures, electrons receive enough thermal energy to rise slightly above the top of the Valence band, effectively elevating the Fermi Level towards the Conduction band (*slightly*).

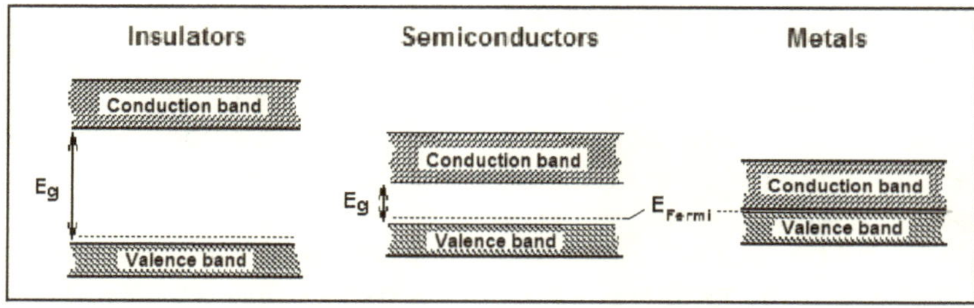

Figure 5.11 Band Gaps for Insulators, Semiconductors and metals.

When a solid forms a well-defined, repetitive, *crystalline* structure, the repetition of the same type of atom in the crystal lattice implies a homogeneity in the solid, giving it characteristic properties unique for each type of crystal. If we take such a crystal and *replace* every third or forth atom in the structure with a different type of atom (known as the "*dopant*" atom) that has one *more* electron than the atoms in the crystal lattice, we effectively introduce one excess electron per dopant atom into the crystal structure that is not needed in the bonds with the neighbor atoms. The accumulative effect of all these extra unbound electrons in the crystal's "Fermi Sea", is to raise the Fermi Level closer to the Conduction band, making that material much more likely to conduct electrons (giving that doped crystal a net *negative* charge).

Similarly, if the dopant atom had one or two *fewer* electrons than the atoms in the lattice, the combined effect would be to give the solid a net *positive* charge, and thus make it a material that attracts electrons. By controlling the concentration, temperature, and pressure of the applied dopant atoms introduced to the crystal lattice (as a hot gas), we are able to change the crystal's band gap and Fermi Level, thereby altering its conductivity as we see fit. Materials whose conductivity is thus carefully engineered are known as "*Semiconductors*".

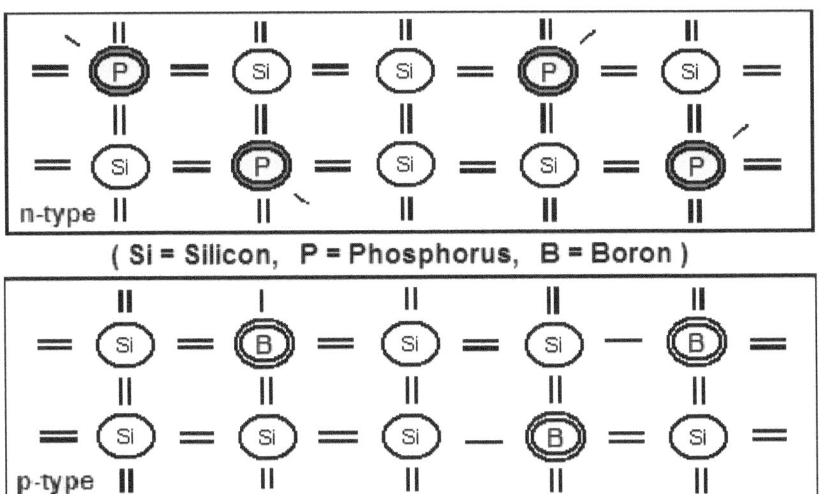

Figure 5.12 Doped substrates forming "n-type" and "p-type" semiconductors.

Referring to Table 4.2 (electronic configurations around atoms), we see that silicon has four electrons in its valence shell ($3s^2$ $3p^2$), implying each silicon atom needs four additional electrons to complete its outer shell (i.e. to reach the magic number of eight). When multiple silicon atoms combine to share their four valence electrons, they develop strong molecular bonds between them (known as "Covalent bonds"), and in the process form a uniform and periodic crystalline structure, where all silicon atoms are equally spaced in well-defined arrays of "rows" and "columns".

If we take such a crystal structure (known as a "substrate") and expose it to a hot dopant gas under pressure (e.g. phosphorus or boron gas) at specific concentration and temperature (e.g. 10^{18} atoms/cm^3 at ~900°C, for ~45 minutes), we find that some percentage of the silicon atoms in the substrate crystal will be supplanted by the dopant atoms. If for example we used phosphorus as our dopant gas, random silicon atoms in the substrate will be replaced by phosphorus atoms.

From table 3.2, we know that phosphorus atoms have *five* electrons in their valence shells ($3s^2\ 3p^3$), which means that *four* of these electrons will bond with the adjacent silicon atoms in the crystal, leaving *one* extra electron that is not involved in the covalent bonding of the crystal substrate. This extra electron thus enlarges the "Fermi Sea", effectively raising the Fermi level in the process. Since these extra electrons are not bound into the crystal lattice, it is very easy to elevate them up into the conduction band, giving this substrate a net *negative* charge, and thereby creating what is known as an "*n-type*" semiconductor.

If on the other hand, the dopant had only *three* electrons in its outer shell (e.g. Boron: $3s^2\ 3p^1$), a net *positive* charge is created in the substrate, creating what is known as "p-type" semiconductor. This doping technique is referred to as "*Chemical Vapor Deposition*" (CVD)[22], and is typically used in conjunction with a masking process known as "*Photolithography*" ("photo" meaning light, "litho" meaning stone, and "graphy" meaning to write; thus literally "to use light to write on stone").

In the Photolithography stage of the process, a thin layer of light-sensitive coating is first spread across the face of the

---

[22] CVD is by no means the only approach to selectively grow "n" and "p" type circuit elements. In fact there are almost as many techniques now as there are letters in the alphabet. Some of these other approaches include Molecular Beam Epitaxy (MBE), Metal-Organic Vapor Phase Epitaxy (MOVPE), Plasma Enhanced CVD (PECVD), Atomic Layer Epitaxy (ALE), (etc.), and a whole host of direct in-situ Ion Implantation and metal sputtering techniques (etc).

substrate. This is then exposed to Ultra Violet (UV) light through an inverse photographic image (similar to a "negative" used in older film type cameras) of a desired circuit pattern (known as a "mask"). Any open areas in the mask allow the UV light to fall on the coated substrate, promoting a chemical transformation in the coating which hardens it to form a temporary gas barrier over that area of the substrate. Areas where the UV light was masked out are not hardened. After the UV exposure, the material is washed in a solvent bath to remove all of the non-hardened coating. Following this, the whole substrate is placed in a CVD oven and exposed to various dopant gases (e.g. Boron, or Phosphorus, etc.) at specific temperatures and concentrations, thereby creating "p-type" or "n-type" pockets as needed in the exposed sections of the substrate.

Following this, the mask is etched away, and the whole masking/CVD process is repeated multiple times over the previous layer(s), building up alternating layers of "n-" and "p-" type components (e.g. p-n junction diodes, Bi-Juntion Transistors, Field Effect Transistors, VCSEL laser diodes, etc.) next to each other as needed to create very complex, multi-layered and densely packed circuits on an extremely small piece of substrate.[23]

Typically hundreds of these identical circuits are fabricated side-by-side on a single disk ("wafer"). At the end of the CVD process, each of these individual circuits are then

---

[23] We note that since most of the optical devices made to filter, tune or otherwise resonate at some desired wavelength need to be fabricated on a scale comparable to their wavelength (typically in nanometers), a fabrication technique with this kind of extreme resolution and control is perfectly suited for the task. By marrying these two disciplines, we are able to design and build *en masse* a variety of very complex filtering / tuning devices using materials such as Silicon and Galium-Arsenide, with exceptional resolution and quality, that only a submicron fabrication technique can produce. This refined technology has already produced a number of innovative solutions with dramatic impact in a number of fields. One such example is included at the end of this chapter under the topic of VCSELs, while others will be encountered in Chapters 7 and 8.

tested, cut from the wafer and mounted individually into ceramic or epoxy packages, and distributed for sell.

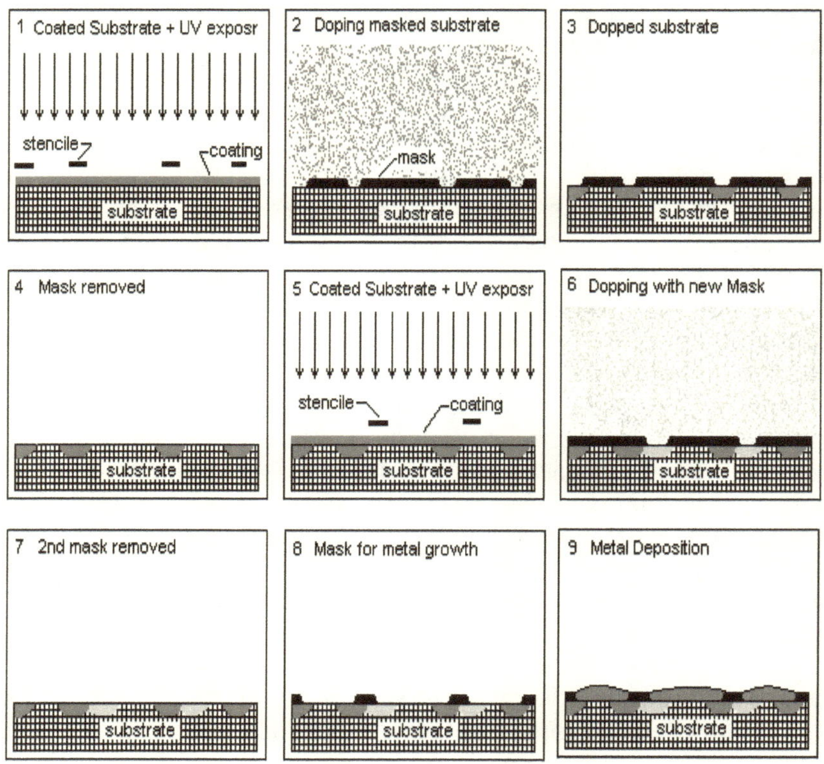

Figure 5.13 Typical semiconductor CVD doping sequence.

When a "p-type" and an "n-type" segment are fabricated adjacent to each other, a "pn-junction Diode" is formed. At the immediate interface between these two segments, excess electrons quickly migrate out of the n-type region into the p-type material, establishing a neutrally charged region at this interface with a relatively high resistance (i.e. this region acts as a temporary insulator), and consequently no further flow of electrons takes place through this junction region.

If we apply an electric field (i.e. a voltage) across this diode, free electrons in the n-type and free "holes" in the p-type (all referred generically as "charge carriers") migrate in response to the field. If the field is applied such that the charge carriers move away from the ends towards the middle of the diode (i.e. "Forward Biased"), the neutral "depletion" region in the middle shrinks (i.e. charge carriers begin to invade the region, narrowing it in the process). If we apply a high enough potential across this junction (e.g. ~0.65 volts for silicon diode), the depletion region shrinks completely (i.e. we exert enough Coulombic force on the charge carriers to force them across the region to the point where they begin to meet in the middle and recombine). At this point, the opposition to current by the junction drops dramatically with a corresponding significant rise in current flow (see Figure 5.14).

Since an electron seeks to fall into one of the available "holes" (attractive vacancies) in any p-type material, a vacant hole obviously constitutes a higher energy configuration than what exists after they combine (analogous to our tank full of water sitting atop a hill, or an excited electron in an isolated atom seeking to fall to a lower energy configuration, etc.).

When we discussed the case of the excited atom in a gas, we mentioned that as the electron falls from a higher orbit into a lower orbit, it expels energy in the form of a photon (see discussion in Chapter 3). Similarly, as an electron from the n-type material falls into a "hole" in the p-type material, it expels excess energy. The amount of energy it expels depends on the composition of the semiconductor. In the case of the material used to fabricate Light Emitting Diodes ("LEDs"), e.g. Gallium Arsenide, the band gap energy happens to correspond to a visible (or in some cases Infrared) wavelength, and hence a visible (or IR) photon is expelled as the charge carriers recombine (as electrons from the n-type material fall into a vacancy created in the p-type material).

Though this photon generation process sounds promising as a possible laser device, we note that simply expelling a photon when fed a little current does not constitute lasing. For any material to function as a laser it must at the very least, undergo a process where ejected photons stimulate the release of other photons (recall that this stimulated emission process is essential if light amplification is to occur, and if the emitted light is to be coherent).

pn Junction Diode

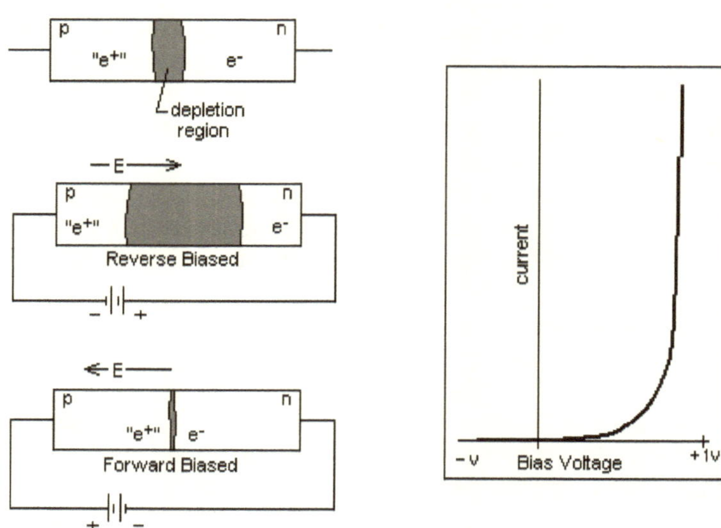

Figure 5.14 Diode profile and current curve.

Several different types of semiconductor lasers have been developed which meet these criteria by effectively allowing released photons to *stimulate recombination* of charge carriers (effectively stimulating a photon release). The size of this lasing cavity therefore is effectively the area where recombination occurs, and is typically around 1×10^{-6}m (the "diffusion length" of the charge carriers. The effective recombination area is a function of the doping concentrations and the type of material used as the substrate). In the basic laser diode, the recombination area is very wide along the interface between the two semiconductor types, but typically not very tall.

As a result, the laser beam emerging from this type of semiconductor laser tends to be compressed along the "vertical" axis and wide along the "horizontal" axis, making it "short and fat" at the aperture. One of the consequences of forming such a beam confined along one dimension (here the vertical axis) is that the beam tends to have a high divergence along that axis. This can be understood using basic Fourier principles, which tells us that the more we try to confine energy along one axis, the more it spreads along that axis as it propagates (this is very analogous to a similar spreading in time that occurs when we compress energy into very short bursts; see example transforms in Appendix B). Therefore, since the lasing region is so extremely compressed along this one axis, we expect the beam to diverge at a different rate along this axis compared to the other axes (see Figure 5.15).

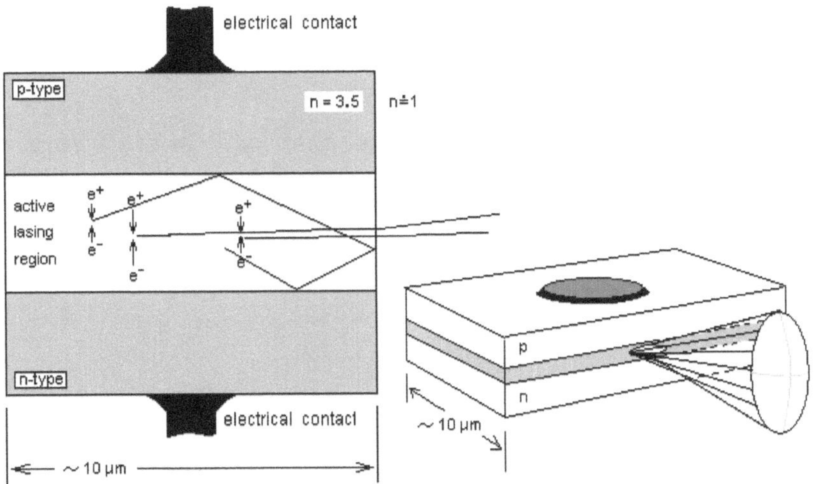

Figure 5.15 Cross-section of a semiconductor laser diode, showing both Total Internal Reflection for photons striking the semiconductor-air interface at high angles of incidence, as well as spatial divergence along the compressed vertical axis.

Another aspect of the semiconductor laser source relates to the cavity of the laser. As we will discuss in the following chapter, most normal laser cavities tend to be constructed by placing a reflective surface (e.g. mirror) at either end to confine the photons within the gain media until we choose to release them. However, in the case of semiconductor diode lasers, the mirror surface is created by the substrate-air interface, due to "*Total Internal Reflection*".

Total Internal Reflection occurs when the index of refraction[24] inside the "source" material is greater than the index of refraction of the external material (e.g. air) and the angle of incidence of the light ray exceeds a threshold value. This critical threshold angle can be found using Snell's law of refraction:

$$n_1 \sin(\theta_1) = n_2 \sin(\theta_2) \qquad \text{Snell's Law}$$

If $n_2 > n_1$ (e.g. region "1" is air), by rearranging terms, we can define the critical angle:

$$\sin(\theta_1) = (n_2 / n_1) * \sin(\theta_2)$$

Since $\sin(\theta_1)$ can never be greater than one, and the ratio of n_2/n_1 is greater than one by definition, $\sin(\theta_2)$ must be the inverse of n_2/n_1 for their product to equate to one at the critical angle:

$$\sin(\theta_2) = (n_1 / n_2) \qquad \text{at } \theta_{critical}$$

Therefore:

$$\theta_{critical} = \arcsin(n_1 / n_2)$$

In other words, θ_1 cannot be a real angle if "$(n_2/n_1) * \sin(\theta_2)$" is ever greater than one (i.e. if $n_1 < n_2 \, Sin(\theta_2)$). However as θ_2 becomes larger, we reach a point where $(n_1/n_2)*\sin(\theta_2)$ is greater than one. At that point Total Internal Reflection occurs and 100% of the light is reflected back. This same effect can be seen by swimming underwater in a pool while looking up towards the surface (see Figure 5.16). Looking directly overhead, you will be able to see objects outside of the water. However, as you look at progressively shallower angles, you eventually reach an angle where the surface of the water undergoes Total Internal Reflection. At that point, the surface

---

24 The Index of Refraction "n" is simply the ratio of the speed of light in a vacuum (300×10^6 m/s) verses the speed at which light propagates in this particular media: $n = c / v$.

interface becomes mirror-like and begins to only reflect light from within the pool, rather than allowing light to transmit across the air-water interface.

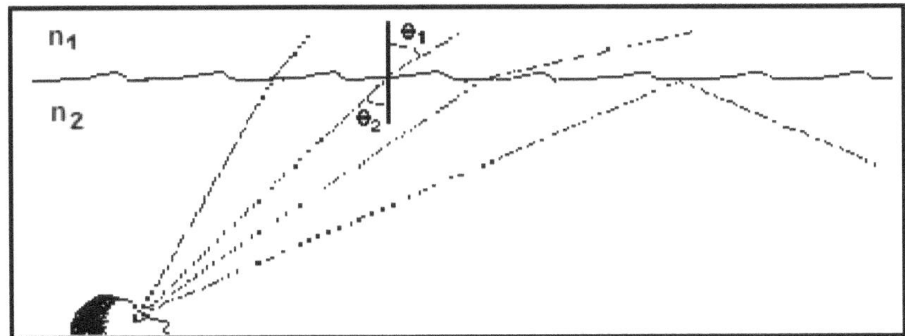

Figure 5.16 example of Total Internal Reflection inside a pool.

Since the index of refraction of the semiconductor material used is so much higher than air, Total Internal Refraction occurs at all angles greater than roughly 20-30 degrees (depending on the substrate's index of refraction). As a result, when the ends of a diode are cleaved and polished, the flat semiconductor-air interface of the surface creates a highly effective mirror for all photons approaching the interface at angles greater than the critical angle, reflecting them back into the laser cavity to stimulate additional recombinations. This effect can even be enhanced by applying a reflective coating to the surface of the diode after polishing.

Exercise: Gallium Arsenide (GaAs) diode laser.

Assuming we have a laser fabricated out of GaAs with an index of refraction $n_2 = 3.6$, radiating into air ($n_1 = 1.0$), find the critical angle $\theta_{critical}$ at which total internal reflection occurs. ($\theta_{crit} =$ 16.1)

Exercise: refracted angles.

Using the above diode laser, find the refractive angle of the emerging beam for the following internal angles of incidence and sketch the emerging beams:

10°, 15°, 17°, 20°

Though these simple pn-junction diode lasers are functional, they do tend to have a few drawbacks, including high drive current (~1kA/cm^2) and a non-uniform beam diameter/divergence (see Figure 5.15). In addition to these "annoyances", there are two very significant problems which degrade the effectiveness of this type of laser, specifically:

1) When the carriers enter opposite type regions (e.g. electrons enter p-type material), they tend to disperse randomly across a fairly wide area. This effectively lowers the recombination density, and hence the gain (i.e. the more you spread out the recombination sites, the lower the probability of a stray photon encountering, and thus, stimulating a recombination).

2) We also find in this type of laser diode that once photons are emitted, virtually nothing acts to keep these photons from migrating into adjacent regions in the semiconductor. As a result, many of the photons released end up migrating into either the p-type or the n-type region well outside the recombination area, and hence their energy ends up being lost into the substrate. We can compensate for this decrease in output power somewhat by increasing the size of the laser cavity (i.e. the recombination area), but that results in an increase in the number of resonant modes present in the cavity (see discussion below).

The latter two problems can be partially resolved through a little creative chemistry, namely, by using a "mix-and-match" approach for the type of materials used for the substrate. The diode created from such mixed semiconductors is referred to as a "Hetero-junction" diode (as opposed to the previously described "Homogeneous-junction" diode). A common substrate for Heter-junction diodes involves GaAs combined with differing amounts of Aluminum doping of Gallium Arsenide (GaAs-Al). The presence of the Aluminum effectively shifts the band-gap, creating new and interesting characteristics for the semiconductor junction. Though this alchemy solution may seem like an obvious approach, in reality, mixing substrates usually creates more problems than it solves, since the dimensions of the respective crystal lattice in each substrate usually are very different (i.e. spacing between each atom is different and hence the layers of each crystal are different for each type of substrate compound), creating large cumulative strains on the structure.

Fortunately however, GaAs and AlAs crystals have nearly identical lattice characteristics, which allows the two to be grown together and/or layered as needed. As the amount of Aluminum *increases* in the mixture, the band gap *enlarges*, and the index of refraction *decreases*. This dual effect has a profound affect on the laser's behavior.

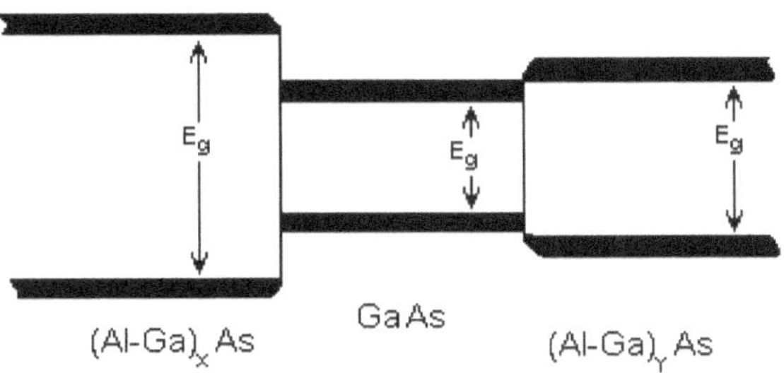

Figure 5.17 Hetero-junction energy levels with various concentrations of Aluminum.

By increasing the band gap, we can improve the confinement of the carriers to the active region, while the decreased index of refraction tends to increase reflections at the edge of the active region, throwing much of the photon energy back into the mix rather than allowing it to migrate uselessly out of the lasing region and back into the substrate.

VCSELs:

In addition to the edge-emitting pn-junction diode laser, a related device known as a "VCSEL" ("Vertical Cavity Surface-Emitting laser", pronounced "vick-cell") has emerged from research in the past twenty years which offers significant advantages over the edge-emitting pn-junction diode just described. These advantages include improved cavity fabrication options, and a significant reduction in fabrication and testing costs, which translates into a much cheaper production/market cost.

One of the drawbacks to the edge-emitting diode laser, is that it must be cleaved and extracted from the wafer it was fabricated in, in order to create the laser cavity edges (since the cavity edges serve as the mirrors) *before* it can be mounted and then tested. Only after performing this labor-intensive process on *each* individual diode, can the manufacturer determine if the individual device is functional or defective (recall that the manufacturing process used in semiconductor device fabrication lacks any fine resolution control on individual atoms, making production yields something of a statistical "roll of the dice").

The advent of the VCSEL eliminates this problem by creating a laser which radiates *vertically* upwards out of the face of the semiconductor wafer. This not only *dramatically* simplifies the testing and production process, at the same time, it provides an easier means to enhance the cavity structure.

Since each VCSEL radiates upwards out the face of the semiconductor wafer (rather than horizontally through the wafer as with the edge-emitting laser diode), it can be tested while still in the wafer, eliminating the need to first cleave, then polish and package each device before they can be tested. Instead, each individual VCSEL is tested by simply attaching micro probes to the electrode on the surface of the wafer and monitoring the individual responses of each unit. This advantage provides an enormous cost savings over the standard edge-emitting pn-junction diode fabrication approach through reduced testing and production costs.

In addition to the significant reductions in costs, the vertical layered semiconductor fabrication process allows us to fabricate almost an unlimited variety of cavity shapes, sizes and structures above the active region. As a result, the performance of a VCSEL can be custom designed with enhancing mirrors, tuning gratings, lenses, etc. (see Chapter 7), within the substrate itself, further enhancing the versatility, performance and cost savings of the technology. Combined, these benefits result in a cost that is roughly one quarter the cost of the edge-emitting diode lasers.

What's more, the vertical fabrication of the VCSEL unit does not impose the narrow height restrictions on the output face of the beam as with the edge-emitting pn-junction diode. This in turn greatly reduces the beam divergence as it emerges from the active region, while at the same time allowing a much easier connection to a fiber-optic cable, thereby eliminating much of the added optics interface needed for the edge-emitting pn-junction diode. And if that wasn't impressive enough, these advantages improve the lasing efficiency to such an extent as to result in a significantly lower drive current, making them cheaper to operate in the field.

Another limitation edge-emitting diode lasers suffer, is the low reflectivity of the mirrors forming the laser cavity itself. This results in a lower operating efficiency and a lower output power from the device. To compensate, many designers of the edge-emitting laser enlarge the recombination region (the active cavity in the diode) to increase gain. Unfortunately, increasing the cavity size increases the number of modes generated by the device (see Chapter 7), which results in an increase in dispersion (pulse "smearing"), limiting the overall length of the fiber optic link in which they can be used (see Chapter 8). Filters can be added to eliminate some of these unwanted modes, but in so doing, we lose a portion of the energy generated by the device, lowering the overall efficiency of the laser.

There are a number of different VCSEL structure types, each with its own particular advantage. However, all of these share some basic commonalities, including a cavity structure grown vertically out the face of the semiconductor wafer, an active region buried in or below the VCSEL cavity structure (which radiates vertically up through the substrate/cavity), as well as a direct electrode contact. The VCSEL cavity structure is typically cylindrical and is placed above the active region, thereby enabling it to channel and refine the photons as they exit the device.

Because the VCSEL beam emerges out the top of the substrate, it is much easier to alter the vertical cavity structure (as well as the substrate of the laser), offering a number of enhancement options to the design of the device with very little difficulty. This at the same time greatly improves the reflectivity and efficiency of the cavity structure, negating the need to enlarge the cavity to compensate for poor output coupling. This smaller cavity size helps minimize the number of modes created in the laser, greatly reducing the effects of "Modal" and "Chromatic Dispersion", allowing the VCSEL to be used on longer fiber optic links, while at the same time increasing the output power and efficiency of the device.

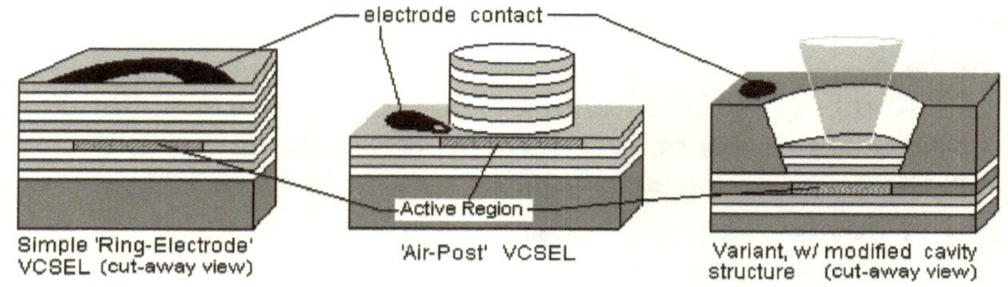

Figure 5.18 "Ring-electrode", "Air-post" and variant structure VCSEL.

Though still somewhat new and maturing, many of the typical wavelengths used in fiber telecommunications are now available in the VCSEL laser (e.g. 850nm, 1310nm etc.).

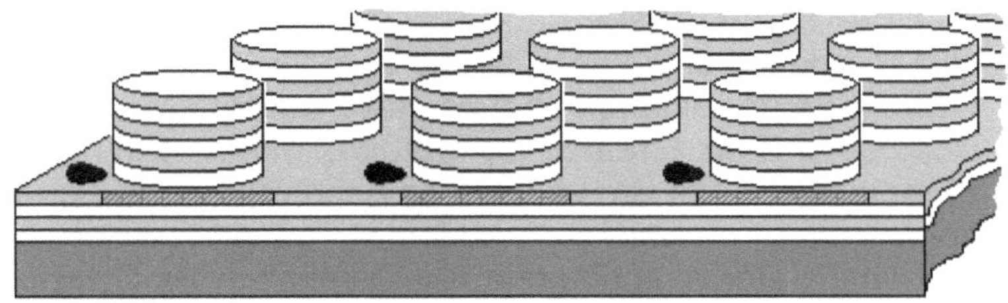

Figure 5.19 An array of "Air-post" VCSELs

The simplest cavity structure design is the "ring electrode", which has the active layer buried under several alternating layers of transparent substrate, capped by a circular electrode ring. Another common design is the "air-post" structure which is slightly more complex, in that all of the substrate grown above and just outside a vertical column immediately above the active region is etched away, leaving what looks like a vertical column of stacked disks. Confinement of the released photons is achieved via Total Internal Reflection at the substrate-air interface surrounding the vertical column.

Laser Line widths:

Before we leave this chapter and move on to the topic of physical components, we need to introduce some of the practical realities that affect the laser emissions, particularly in dealing with "spectral purity" and "spectral broadening".

Beginning with our Quantum description of the material at the heart of the laser (atoms and molecules), we have thus far assumed an *ideal* output from the laser, particularly with the emissions being assumed perfectly monochromatic (i.e. an infinitely narrow line width with one and only one wavelength, e.g. our HeNe laser wavelength being exactly 632.8000000... nm). However, in light of what we learned about the fundamental aspects of the Quantum world, Heisenberg's Uncertainty Principle in particular, it is highly unlikely that we would get such a perfectly monochromatic output from any generator (be that a crystal oscillator in a radio, or an atomic resonator composed of vibrating atoms or molecules, etc.).

We know, for example that we have absolutely no control over *Doppler shifting* (created by random movement of each atom or molecule). Nor do we have any control over the effects of *Spontaneous emissions* (randomly generated photons produced when atoms spontaneously de-excite), or *Collisional Broadening* (the disruption of the atomic oscillations by random collisions with neighbor atoms), etc. To refine the scope of our discussion into a slightly more *realistic* perspective, we will need to consider the impact each of these effects have on the output energy of our laser.

Doppler Spreading:

Anytime a source is in motion relative to an observer, be that involving a moving train horn, a spiral galaxy, an atom moving through a hot gas, etc., the observer will detect a shift in the wavelength of the energy being radiated by the source. This quite puzzling phenomenon became something of a hot topic of discussion only after man began to travel by machines capable of exceeding 30 – 40 mph, where the effect becomes noticeable. Johanne Christian Doppler was the first to quantify this effect mathematically (ca.1840), as he described his experiment involving a group of blaring trumpeters, a moving steam train, and a stationary observer:

$$F_{obs} = F_{src} * (v + V) / v \qquad \text{Doppler compressed}$$

$$F_{obs} = F_{src} * (v - V) / v \qquad \text{Doppler stretched}$$

where "V" = relative speed between source and observer, "v" = speed of propagation which is ~345 m/s for sound in air (or "c" = 300×10^6 m/s for E&M waves in free space, etc.). Since our ears and brain interpret "pitch" or frequency/wavelength as a function of pulses on the eardrum (i.e. the number of percussion wavefronts hitting the eardrum per second), any movement of the source alters the time between each successive wavefront detected by an observer that is not moving with the source, thereby altering the perceived frequency at the observer.

If the distance between source and observer is decreasing (e.g. an approaching train), the observer will detect more percussion wavefronts arriving per second than does an observer traveling with the source (or an observer stationed well behind the source). To that forward observer the "pitch" (frequency) of the oncoming train horn appears to be higher (i.e. shorter time between waves, and hence a higher frequency) up until the point where train passes the observer. The instant the train passes the observer, the signal is neither compressed nor stretched for a brief moment and it therefore sounds the same to this observer as it does to the passengers on the train. When the train passes however, it is now receding away from the observer and as a result, it takes longer for each successive wavefront to arrive at the stationary observer, and hence the "pitch" seems lower (i.e. longer wavelengths, and thus lower in frequency).

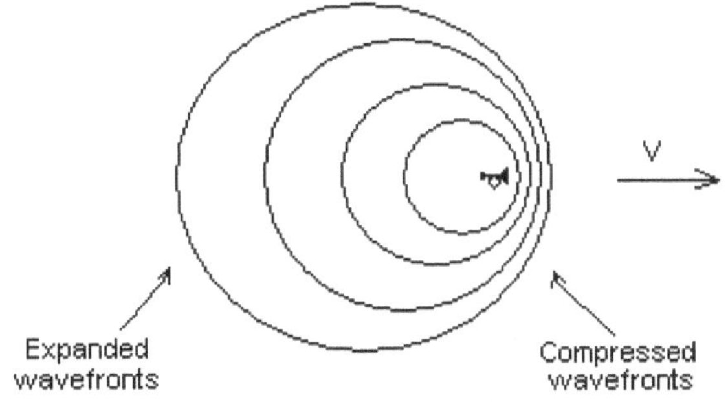

Figure 5.20 Doppler effect on propagating wavefronts.

This same affect occurs with *any* energy that travels as a wave, including radio waves (e.g. Doppler broadening can affect the bit error rate for cell phones traveling on high-speed trains), and light waves (e.g. the "red-shift" and "blue-shift" in light coming to us from distant spinning galaxies, allowing us to determine the direction of spin of these galaxies), etc.

In the case of light generated within a laser cavity, If the atoms or molecules in the gain media are all moving at considerable speeds and in random directions within the cavity, they each produce a slightly different Doppler shifted wavelength. Consequently, even if they were all triggered by a photon at the desired wavelength, their random motion creates a *range* of wavelengths in the light exiting the laser cavity. This *Doppler Broadened* output appears as a broadened spectral peak around the desired output wavelength.

Using Planck's model for the "Black body radiator", we can write the fraction of atoms or molecules that have speeds between "v" and "v + dv" as:

$$df(v) = SQRT[\, M / (2 \pi kT) \,] \; \exp(Mv^2 / 2kT) \; dv$$

where "M" = mass of the individual media atom or molecule (in kg), "k" = Boltzmann's constant (1.380×10^{-23} J / Kelvin) and "T" = Temperature in Kelvin.

The Doppler line shape function is similar:

$$dS(f) = SQRT[\, M c^2 / (2 \pi k T F_o^2) \,] \; \exp[M(F-F_o)^2 / 2kT] \; dv$$

It can be shown [6] that this suggests a Doppler line width of:

$$\delta_D = 2 \; SQRT(\, \ln(2) \; 2RT / M \,) \; / \lambda$$

(where R is the universal gas constant, 8.314 J/Kelvin). Reducing the known constants in the equation (and writing our mass term in grams "m"), we find the ratio of the Doppler line width to peak frequency is:

$$\delta_D / F_o = 7.16 \times 10^{-7} \; SQRT(\, T / m \,)$$

Exercise: Doppler broadening.

Find the Doppler line width of our HeNe laser at room temperature, λ=632.8nm, and m=20.2 g/mole. (1.3 GHz)

<u>Broadening by Spontaneous Emission</u>:

To investigate line broadening due to Spontaneous Emissions, we will consider it in terms of the Heisenberg Uncertainty relation. In addition to the position-momentum Uncertainty relation already discussed, we note that there are several other forms of the Heisenberg Uncertainty relationship that addresses the uncertainty in a number of other "mutually exclusive" (more correctly, "non-commuting") observable quantities. In the case of the purity of the wavelength produced by our laser, we recall that energy and wavelength (or frequency, since $F = c/\lambda$) are related as: Energy = N h F. The form of the Heisenberg relationship that deals directly with energy is:

$$\Delta E \, \Delta t \quad \geq \; h / 2\pi$$

This statement should seem reasonable in light of the position-momentum Uncertainty Relation we have already discussed, and the fact that energy is directly related to frequency (and thus inversely related to time). We know that with sinusoidal waveforms, $F = 1/t$. Therefore, the same relationship holds between the uncertainty in frequency and the uncertainty in time (or wave period): $\Delta F = 1 / \Delta t$. Therefore, if we could somehow narrow the uncertainty in time to zero, we would end up with an infinite uncertainty in frequency.

A similar Uncertainty relation exists for the number of photons in the cavity "N", but we will assume that during *steady state* operation, this uncertainty is much smaller than the uncertainty in the frequency. As a result, $\Delta E = N\,h\,\Delta F$. Plugging this back into our Uncertainty relation, we find:

$$(N\,h\,\Delta F)\,\Delta t \; = \; h / 2\pi$$

or:

$$\Delta F = 1 / (2\pi \, \Delta t \, N)$$

If we assume the velocity of the atoms or molecules within the cavity is relatively small (i.e. the Doppler frequency spreading is small by comparison), the primary cause of the uncertainty in frequency is then the "wildcard" of Spontaneous Emissions (i.e. excited atoms or molecules de-exciting as random, un-stimulated events). Since Spontaneous Emissions are the result of randomly decaying metastable states, and since we typically know the approximate lifetime of the metastable states, we can estimate the line broadening due to Spontaneous Emissions as a function of metastable state lifetime, using the above equation.

Example: Spontaneous decay broadening.

If the lifetime of the metastable state is 10^{-6} seconds, and the number of photons present in the cavity "N" is 10^{15}, estimate the width of the output peak.

$$\Delta F = 1 / (2\pi \, \Delta t \, N)$$

$$\Delta F = 1 / (2\,\pi \;\; 10^{-6} \text{ seconds } \times 10^{15} \text{ photons})$$

$$= 0.16 \text{ nm}$$

<u>Collisional Broadening</u>:

If we follow Max Planck's lead and think of the individual atoms or molecules in the gain media as individual radiating oscillators analogous to the simple Lorentz "ball-and-spring" model, we know that anytime one of these oscillators is struck by another, its oscillation is abruptly halted mid-swing. Immediately after the collision event, it will once again begin to oscillate, but it will lose all "memory" of where it was in its oscillation prior to the collision. As a result of the collision, the wavelength of that oscillator is changed for the brief instant of the collision, producing what is known as "Collisional Broadening". The greater the density of the gain media, the greater the probability of a collision (i.e. the more frequently collisions will occur).

If we know the density of the gain media (i.e. number of atoms/molecules and pressure) and the approximate size (i.e. the "cross section") of an individual atom or molecule, we can calculate the average time between collisions, and from that estimate the amount of wavelength change due to Collision Broadening.

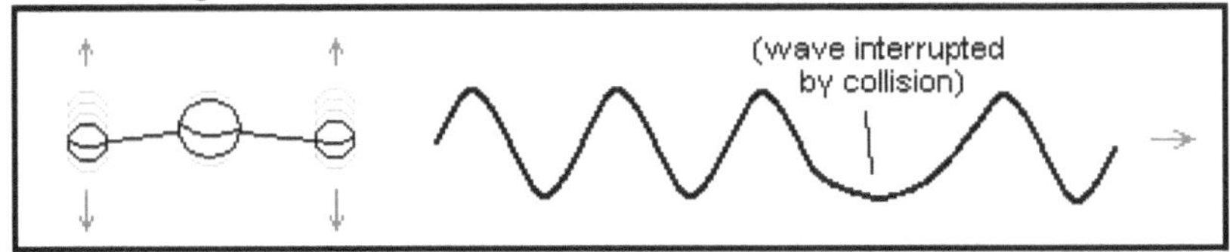

Figure 5.21 Wave emission interrupted and distorted by collision.

If we define "N" as the number of atoms or molecules per cubic meter, "P" as pressure (in Pascals), "T" as temperature (in Kelvin), "t_{coll}" as the time between collisions, "σ" as the "cross section" of an individual atom or molecule, and "V_{ave}" as the average velocity of an individual atom or molecule, we can estimate the amount of collisional broadening as follows:

$$N = \text{density} \left(P_{(pa)} / T_{(k)} \right)$$

$$t_{coll} = 1 / (N \sigma V_{ave})$$

$$\Delta F = 1 / t_{coll}$$

Example: Collisional Broadening in CO_2

Given a density of $73.3 \times 10^{21} / m^3$, a cross section "$\sigma$" = $5\text{x}10^{-19}$ m^2, and an average velocity "V_{ave}" = 500 m/s, find the average time between collisions (t_{coll}) and the Collisional Broadening of a CO_2 gain media at 300 Kelvin and 2 atmospheres ($2\text{x}10^5$ Pascals):

$$N = 73.3 \times 10^{21} / m^3 \ (2\text{x}10^5 \text{ Pa} / 300\text{K})$$

$= \quad 48.9 \times 10^{24}$ molecules

$t_{coll} \quad = \quad 1 / (48.9 \times 10^{24} \times 5 \times 10^{-19} m^2 \times 500 m/s)$
$\quad\quad = \quad 81.8 \times 10^{-12}$ sec

therefore,

$\Delta F \quad = \quad 1 / 81.9 \times 10^{-12}$ sec
$\quad\quad = \quad 12.2 \times 10^9$ Hz

Chapter 5 Review questions:

1. Explain the concept of "pumping the gain media".
2. Name and describe three common methods used to pump the gain media.
3. How does the electron discharge scheme pump a gain media?
4. Name at least four reasons why the energy in the electron discharge does not all end up in the external laser beam (i.e. where else does this energy go)?
5. How does Helium in the HeNe laser media increase the power of the output beam?
6. Why is the $3s_2$ -> $2p_4$ transition in Ne metastable? (Hint: see footnote #18)
7. In our HeNe example, how much energy did we put into the "engine" compared to the final amount that made its way into the laser beam? Where did the rest of the energy go?
8. Why do atoms join together to form molecules?
9. In terms of bond energy, why don't isolated Helium atoms form He_2?
10. What is an "anti-bond" and what affect does it have on a molecule forming?
11 How does the Pauli Exclusion Principle apply to molecular formation?
12. How does knowing the bond strength help us predict the resonant frequency of a molecular bond? What is the relationship between bond strength and the frequency/wavelength of the related spectral peak?

13. How does a molecule being made to vibrate imply storing energy in that molecule?
14. Why is the resonant frequency of an isolated molecular bond different when found in a more complex molecule?
15. How many vibrational modes are there to a CO_2 molecule? Why do we only measure three?
16. What is a degenerate mode?
17. Why are vibrational modes in molecules in the Infrared and electronic resonances (spectral lines) in the visible region?
18. Name the two methods that can be used to drive a molecule into vibration.
19. What happens to the electronic levels as more like-atoms are combined in a solid?
20. What are Forbidden regions? What are band gaps?
21. How do the band gaps differ in Insulators, Semiconductors and metals? Explain in terms of the Pauli Exclusion Principle and Fermions.
22. What is the "Fermi Sea" and the Fermi Level? What does the Fermi Level tell us about conduction properties?
23. What is a dopant, and how does it affect the Fermi Level?
24. How does a dopant alter the electrical characteristics of a crystal lattice?
25. Describe how a semi-conductor device is fabricated.
26. How is a pn-junction diode formed?
27. How does an external electrical potential applied to a diode alter its conductivity?
28. How is a semiconductor laser diode different from an LED?
29. What is Total Internal Reflection, and how does it affect the operation of an edge-emitting laser diode?
30. What is the relation between the laser diode recombination region and the lasing effect?
31. What is a Hetero-junction diode laser, and what advantages does it offer over the Homo-junction laser?
32. What is a VCSEL?
33. Name five advantages VCSEL technology offer over edge-emitting junction diode lasers?
34. Explain Doppler Broadening.

35. How does Spontaneous decay effect the lasing Bandwidth?

36. Explain how atomic / molecular collisions broaden the output wavelength.

37. Following the discussion around Figure 5.5, would we expect Hydrogen and Chlorine to form a diatomic molecule? Would such a molecule have a rotational spectra?

38. If an "H-Cl" molecule were to bond with a bond strength ("k") of 514 N/m, what would be the fundamental vibrational frequency and its wavelength? (λ = 3.34 μM)

39. Given a semiconductor crystal made of GaP with n_2 = 3.45 radiating into air (n_1 = 1.0), find the critical angle $\theta_{critical}$ at which total internal reflection occurs. How does this compare with GaAs? Which has less beam divergence? (GaAs)

40. Given T= 300 Kelvin, P=3x10^5 Pascals, N=85x10^{21}/m^3, a cross section "σ" = 7x10^{-19} m^2 and an average velocity "V_{ave}" = 850 m/s, find the photon ΔF. (50.6 GHz)

Chapter 6: Maxwell's Equations and the Propagation of light

Now that we have a good feel for how coherent monochromatic light is created by a laser, we need to discuss how this energy travels once it leaves the laser and propagates through various media (e.g. air, a fiber optic wave guide, etc.). Since these emissions are Electromagnetic by nature, we will need to explore what Electromagnetic fields are, and how they operate, and to do that, we will need to introduce Maxwell's Equations.

To accommodate those who are either not "well-versed" in *math-speak*, or perhaps just a little "rusty", we will begin this discussion at the "shallow end" of the pool, and over the course of the material gradually transition into a little more depth. Even if it has been a few decades since your last math class, this initial overview of the topic should still offer a great deal of general insights into what "radio waves" are, and where they come from. Again we note that we have made an effort to make the initial sections of this chapter as "user friendly" as possible, and strongly urge the non-math "savvy" reader to at least read through the discussion.

As most "techies" know, radio waves, infrared waves, light waves, x-rays, etc. are all made of the same electromagnetic "*stuff*". The only difference is the frequency at which each oscillates. For example, radio waves oscillate at anywhere from ~10^3 cycles per second (Hertz, or "Hz") to ~10^{10} Hz; infrared waves oscillate at anywhere from ~10^{12} Hz to ~10^{14} Hz. Visible light, being something of a special form of electromagnetic waves (in that our eyes just happened to be tuned to that portion of the E&M spectrum where water has freakishly minimal absorption, known as the "visible spectrum"), comprises a very narrow segment of the electromagnetic spectrum just above ~10^{14} Hz (~400nm to ~700nm). X-rays and gamma rays oscillate at ~10^{16} Hz and higher, and tend to oscillate so rapidly and with so much energy, that they penetrate through all but the *densest* materials with relative ease (e.g. flesh, bone, wood, stone, metal, mathematical equations, etc.).

The ElectroMagnetic Spectrum

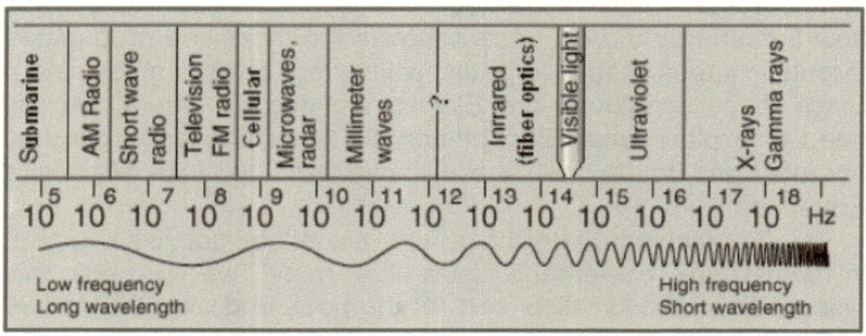

A simple example of an electro*static* field can be demonstrated by taking a balloon and rubbing it on your hair (or a glass rod rubbed through some dry fur, or by walking across a dry nylon carpet, etc.). The process of rubbing displaces electrons through friction from the hair (fur, carpet, etc.) onto the balloon (or glass rod, etc.). Since the balloon is made of a non-conductive material (e.g. latex), the electrons accumulated on it can't migrate away from each other. As a result, they form a concentration of excess charge on the balloon at the point of contact. If you then move that area of the balloon *close* to a "test" pile of dry lint (or paper clippings, or clean dry hair, etc.), the test pile will at some point begin to move towards the charges well *prior to making contact* with the balloon.

A similar effect can be demonstrated using a magnetic field, by placing a magnet *near* a small piece of iron or tin (e.g. a paper clip). At some point *prior to making contact*, the small metal object will similarly begin to move towards the magnet even though no physical contact is made. How does the balloon or magnet apply a force on the test objects, when *no physical contact* exists to transfer the force between them?

We explain this "action at a distance" by saying that some sort of energy "field" clearly radiates out from the charges on the balloon and magnet through empty space with sufficient strength to move the test objects. In fact, since gravity is pulling these test objects down, we conclude from the fact that they jump to the charged balloon and magnet, that these electric and magnetic fields are actually *stronger* at this distance than is the force of gravity acting on the test objects.

Though most people are at least somewhat familiar with the phenomenon of light and electromagnetic wave propagation from everyday experiences (e.g. sunlight, TV signals, cell phones, etc.), most tend to view them as something of a vague, ethereal mystery. Most of this mystery lies in the fact that the vast majority of these fields are completely invisible to us, making our senses almost useless when it comes to detecting and exploring these fields. In a sense, this is the same problem we had when trying to study the nature of electrons and atoms. Consequently we find ourselves once again having to analyze these things using some form of mathematical models and/or tools.

<u>Basic "pre-flight"</u>:

Before we plunge into the deep end here, let's first talk a little about the various bits of "shorthand" we're going to be using. As with any topic or field of study, "80%" of the challenge is just "learning the lingo", or in this case the symbols used to express certain concepts or represent certain "operations" to be performed.

If you think about it, most of the symbols we use every day have no intrinsic meaning of their own (which includes of course, all the letters on this page). However after a little time under the tutelage of your grade school/primary school *drill sergeant* of a teacher, you were slowly conditioned to automatically associate such symbols with certain specific meanings. For example:

K, Ch, ⬣, ∇, ⊗, ♪, $, £, √‾, Δ, 中人

have no intrinsic underlying meaning embedded in the fabric of the universe or otherwise imprinted on our DNA. Yet when familiar symbols such as these cross our gaze, they automatically invoke an almost reflex response in us, simply because we learned over the years to associate their compact symbology with the specific meaning they are intended to represent.

One of the biggest barriers most people have in fact when learning a new subject, is an almost *gut* level reflex akin to *fear* at seeing notation that is an almost alien thing to them. Yet after they learn the meaning that is associated with those symbols, they are hard pressed to remember what all the crying was about.

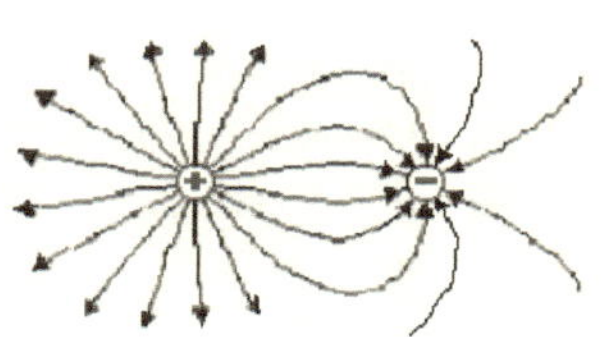

The static electric field around two electric charges.

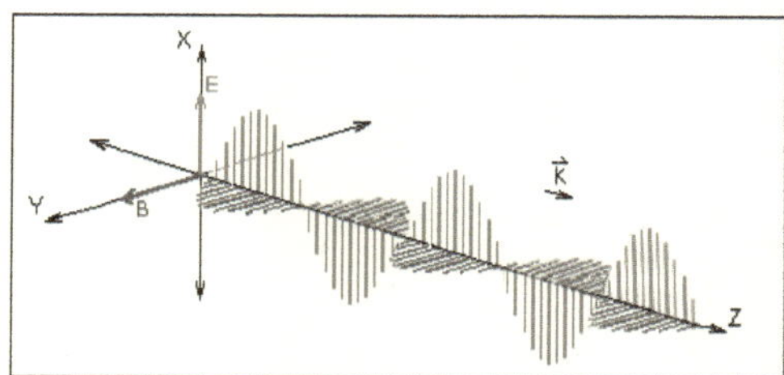

The sinusodial flow of an electric (E) and a magnetic (B) field along the 'z' axis.

For example, "**E**(x,y,z, t)" doesn't exactly roll off the tongue. However once we learn this is a shorthand notation for the "Electric field" (e.g. from our charged balloon, etc.) which flows out into space (x,y,z) over time (t), this symbology starts to develop a bit more meaning for us.

Similarly, "**B**(x,y,z, t)" represents the "Magnetic field" (e.g. in our previous magnet/compass experiment). Any time electric charges are put in motion (i.e. an electric current), we then detect a Magnetic field around those moving charges. Since the humble little electric charge, "q", is the source of both types of fields, it should come as no surprise that the Magnetic field is directly related to the Electric field. You can *not* have one without the other[25].

Any time you have an accumulation of such electric charges (say after rubbing your nylon jacket over your dry hair, or after walking across a dry nylon carpet and reaching for that metal door knob, etc.), those charges radiate an Electric field out into space. This Electric field manifests itself anytime those accumulated charges are brought near other charges around them, causing the charges around them to move without any direct contact (see our "Cell Phones 101" tutorial on EpiphanyBySteveLee.com, misc. tab for more extensive discussion on demonstrating this "action at a distance" effect, as well as on how we use these Electro-Magnetic fields to "telecommunicate").

Anytime we study charges at rest, we are talking about *Electro-Statics.* When we put those charges in motion, we are then talking about *Electro-Dynamics.* The biggest difference between the two is in the fact that when a charge is stationary, we (or more correctly, other charges around it) sense only a static *Electric field* surrounding that charge. When we then put the charges into motion however, we then detect both the *Electric field* and a *Magnetic field.* It is when charges are put into motion and both the Electric and the Magnetic fields are detected, that we are able to use them to send "encoded" radio waves out across great distances (e.g. broadcast radio, television, cell phones, radio signals to NASA deep space probes, E.T. Phoning home, etc.).

Since these fields distribute themselves out into space, we obviously need to be able to describe how these fields propagate across the 3-dimensions of real space (plus time of course). Hence the "x,y,z, t" part of our field notations listed above. And since these fields are 3-dimensional (plus time), in order to understand what these fields are doing, we need some way to analyze their shape and distribution in space and time (e.g. how strong are these fields along each axis? Are they stronger in one direction than another? Can we use that

---

25 Some might want to argue you can have an Electric field and no Magnetic field, if the charges are held stationary. Keep reading.

preferred direction to allow us to "aim" them in a given direction and thus focus energy down a given axis? Etc.).

Before we attempt to describe these fields across a region of space, let us first introduce what is called a "vector". A vector can be thought of as an arrow-like object that describes both the *magnitude* of the field and its *direction*. An example of a vector would be the typical description used to define something's position relative to some reference point (e.g. "Ottawa is 250 miles north by northeast of here", or " the plane is now 20 miles due east of London at 20,000 feet", i.e. both magnitude *and* direction).

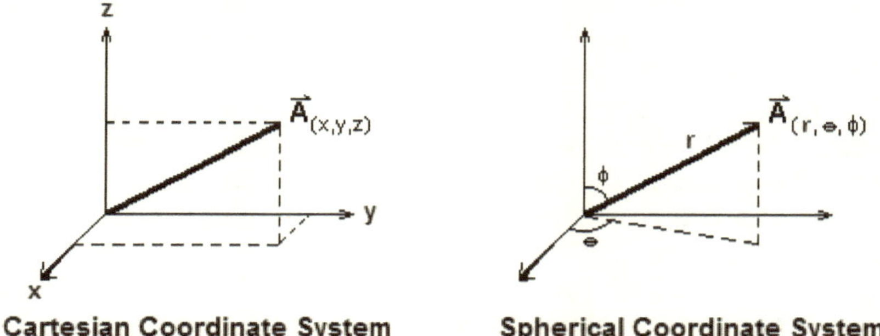

Figure 6.1 Examples of vectors described in two possible coordinate system reference frames.

When dealing with vectors, we find it helpful to decompose them into components related to the reference frame we are using. For example, if we are using an "x,y,z" Cartesian coordinate system, we can describe a vector by saying it is composed of some amount along the x-axis (the "x" component), some amount along the y-axis (the "y" component), and some amount along the z-axis (the "z" component).

In other words, we can decompose a vector into an ordered set of individual components, and examine each component separately. We can decompose vectors into components due to the fact that each of our coordinate axes is perpendicular ("orthogonal") to the others. This allows us to treat each component independently, since motion along one orthogonal axis does not affect motion along the others. Being able to decompose vector quantities (e.g. fields, forces, etc.) in this way, dramatically simplifies any problem involving vectors (including discussions involving electromagnetic fields, as well as describing the location of an electron orbiting an atom, etc.), allowing us to break down a complex 3-dimensional problem into three separate smaller problems such as: **E(x,y,z)** = (E_x, E_y, E_z), vastly simplifying our analysis. If a given vector happens to be aligned entirely along one axis, e.g. the "y" axis (making the "x" and "z" components zero), then: **E(x,y,z)** = (0, E_y, 0); if the vector lies in a plane (e.g. the x-y plane), it would only have "x" and "y" components: **E(x,y,z)** = (E_x, E_y, 0); etc.

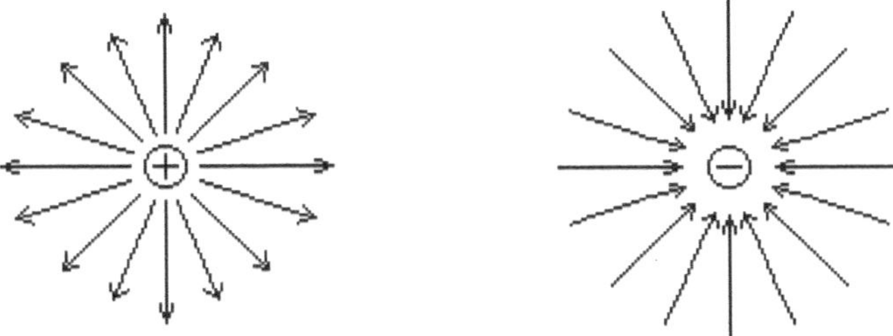

Figure 6.2 Electric fields radiating from stationary charged particles

As for the Electromagnetic fields themselves, we reiterate that the *source* of these fields is the humble little electric charge (individual charges, or a collection of charges as a group), acting largely in keeping with the basic rules that:

- "*Opposite* charges *attract*"
- "*Like* charges *repel*"

It is this attraction and repulsion between charges that is "communicated" through these electromagnetic fields.

The strength of the Electric field radiating out from a charged particle is defined using Coulomb's Law:

$$E(r) = q / (4 \pi \varepsilon r^2) \qquad \text{Coulomb's Law} \qquad (6.1)$$

where "q" is the net charge, "ε" is the "permittivity" of the material the field is moving through (i.e. how the atoms and molecules in that material affect the field), and "r" is the distance separating the charge and the point of interest[26] (e.g. the observer's locations). We note that the electric field naturally radiates out from the point source in a spherically symmetric volume of space, which geometry is the origin of the "4π" in the denominator.

Coulomb's law tells us that the Electric field strength produced by a collection of charges is proportional to the amount of *net* charge "q" on that particle, and inversely related to the distance squared between charges and observer. The more like charges we accumulate on that particle, the stronger the Electro-Static field will be around it. Conversely, the "$1/r^2$" in Coulomb's law tells us that the farther we are from the charges, the weaker the electric field is around us. This law also indicates that the strength of this field is affected by the permittivity (ε) of the material around the charge, be that air, water, plastic, glass, etc. (see discussion below).

As mentioned, anytime a charge begins to *move* relative to the "observer" (i.e. is accelerated from rest), we know from experience that we then begin to detect a *Magnetic* field around the motion of that charge, to which other charges react (one can verify this simply by wrapping a wire around a toy compass, and running a current through the wire). Typically, this Magnetic field is *much* weaker than the Electric field at normal velocities ($V << c$, where "c" is the speed of light, 300×10^6 meters per second), as we will discover shortly.

---

26 Note that at r=0 (the center of our test charge), Coulomb's law suggests that the Electric field is infinite. Since an "infinite" field necessitates an infinite amount of energy (suggesting a "zero point" energy source), this conclusion remains a lively topic of debate in both the study of Electrodynamics as well as Quantum Mechanics.

The next equation we would like to discuss is the "Lorentz Force equation", which describes the amount of *force* a charged particle experiences when exposed to either of these two fields. As mentioned in Chapter 2, the Electric force component due to the *Electric field* is directly proportional to the amount of charge on our particle and the strength of the external Electric field the charge experiences. The Magnetic force component on our test charge due to the *Magnetic field* also depends on both of these factors, as well as the *relative* velocity of our charge. This implies that stationary (or slow moving charges) generate a very weak Magnetic field, while fast moving charges generate a much stronger Magnetic field. (Note that since motion is relative, we could detect the same effect if either our test particle was stationary and we quickly moved a Magnetic field past it, or the source charges were stationary and the test particle moved quickly past the source.)

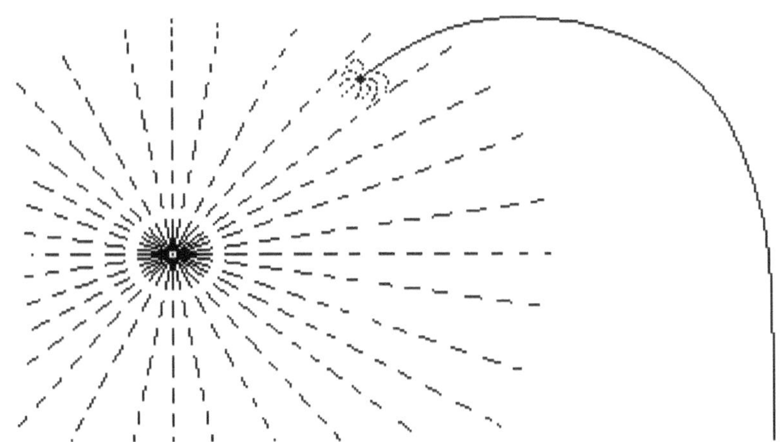

Figure 6.3 Charged particle caught in a field

Further, the force component from the Magnetic field is defined to be the "cross" product of the two vectors (**v** x **B**), which implies that the deflecting force on our test charge by the Magnetic field will be *perpendicular* to the direction it is traveling.

We observe this behavior when charged particles move through a Magnetic field in a Cloud Chamber, *spiraling* as they go).

$$\mathbf{F} = q\mathbf{E} + q(\mathbf{v} \times \mathbf{B}) \qquad \text{Lorentz's equation} \quad (6.2)$$

where **bold** letters indicate vector quantities (i.e. both magnitude *and* direction): **F** = Force vector, **E** = Electric field vector, and **B** = Magnetic field vector.

Once we have determined the force on our test charge resulting from this external field, we can use that to easily find the acceleration of our test charge under the influence of that force using Newton's Second law, F = ma (i.e. the more massive the particle, the less it will accelerate for a given force).

Though Lorentz's equation tells us what amount of force or deflection we can expect the charge to experience, it does *not* tell us much about the *relation* between the Electric and the Magnetic fields, nor does it provide a way to determine how these fields propagate beyond the region of their source out through any surrounding media.

To gain that level of insight (and a great deal more), we must turn to Maxwell's Equations. A number of scientists have contributed to the understanding of electromagnetic fields, including Gauss, Faraday, and Ampre. However James Clerk Maxwell is attributed with the culminating efforts that tied both the Electric and the Magnetic fields together, placing Maxwell at the forefront of this subject. The seemingly simple little set of equations that collectively bears his name, contain a wealth of information about this very fundamental force of the universe. Since the Electro-Magnetic force represents one fourth of the known forces in the Universe (Gravity, Strong Nuclear and Weak Nuclear forces being the other three), Maxwell's equations provide a very powerful and elegant set of tools to enable us to unlock a vast portion of the mysteries of the Universe.

In fact, the amount of knowledge bound up in their compact little form is so profound in its implications, that it has been said that their insights exceeds the total composite knowledge of man. Whether entirely accurate or something of a *slight* exaggeration, it cannot be denied that since they were codified by Maxwell shortly after the American Civil War, nothing about man's culture has been the same since (well, except perhaps man's own self-aggrandizing nature, which seems to be one of the more immutable constants of the universe – however that's a topic for a few of our other books) (*shameless plug*).

The first known use of Electromagnetic waves to convey information across any significant distance was back in the 1890's when Guglielmo Marconi and Nikola Tesla independently began experimenting with high power alternating electrical currents. In the process they found that if they got the currents to alternate at a high enough frequency (e.g. a million cycles per second) on a long wire or metal structure (the antenna) roughly a quarter wavelength[27] tall, an electronic circuit some distance away tuned to that same frequency would resonate with the energy radiating off the antenna[28]. The first laser on the other hand was not built until the 1950s, both of which we point out, were founded (at least in part) on the *theoretical* principles bound up in Maxwell's Equations.

In a very real sense, Maxwell and his peers used mathematics to enabled them to "divine truth" *decades* before we even had the technical means to explore what their equations suggested, effectively using math to predict and shape the future. It is therefore no exaggeration to say that Maxwell's equations were at the very least, an *incredible* anachronism (particularly when you consider the extreme crudeness of our technological skill in 1865). These four little concise statements, which brought us instant global communications and mass

---

27 One wavelength is the distance between two peaks of the wave as it travels. Therefore, Wavelength = Speed/Frequency = c / F , where "c" is the speed of light in free space, 300 million m/s.

28 Tesla actually went so far as to build a few gigantic towers topped with oblate metal spheres which he powered in an attempt to create a system of such towers to *electrify the atmosphere*. His goal in doing so was to prove that people miles away could pull this energy out of the atmosphere to power light bulbs and other electrical equipment. Fortunately for us, this technique never quite caught on as a power distribution system (due to the extreme loss between transmitter and receiver, limiting the currents delivered), but his efforts did prove it was possible to use this technique to instantaneously transmit information across great distances.

dissemination of information, have so profoundly altered our world and even our very perception of life, that, other than perhaps Einstein's $E=Mc^2$, no other scientific principle has even approached so pronounced a change in man's collective awareness. Some food for thought.

To begin our review of Maxwell's Equations, we need to introduce another symbol that represents a mathematical "tool" that will help us define how a field is distributed across space. This "tool" is known as the "Del operator". When applied to a vector describing one of these fields, the Del Operator[29] "$\nabla(x,y,z)$" effectively finds the 3-dimensional "spatial rate of change" of that field, describing how the distribution and strength of that field changes along each of the three spatial axes (e.g. x,y,z):

For example, if we have a mathematical function that describes a particular mountain, we can apply the Del Operator to that function to determine the quickest way down the mountain (which of course would be very useful, particularly if you had a death wish):

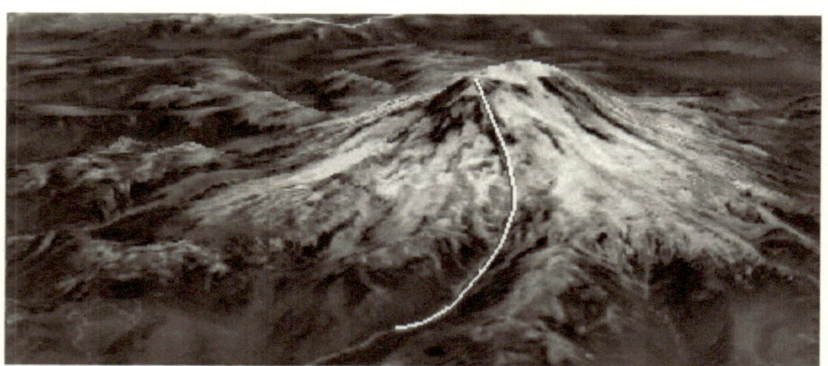

[29] If we pry open the Del Operator to see what's inside, we find: $\nabla(x,y,z) = \hat{\imath}\,\partial_x + \hat{\jmath}\,\partial_y + \hat{k}\,\partial_z$, i.e. a separate "change operator" for each axis, where the "$\hat{\imath}$, $\hat{\jmath}$," etc. represent unit vectors along each of the three axes.

Applying the Del Operator to our fields as shown below gives us the "*divergence*" of these fields:

$\nabla \bullet \mathbf{E}$ (x,y,z, t)
$\nabla \bullet \mathbf{B}$ (x,y,z, t)

In dealing with vector fields, Helmhotz realized that in order to completely define any 3-dimensional field, we need to determine two things about that field:

1) Its "divergence" = $\nabla \bullet \mathbf{F}$ (the total energy radiating out from the field's source), and:

2) Its "curl" = $\nabla \times \mathbf{F}$ (also known as the field's "rotation", i.e. how that field rotates or curls as it distributes itself out into space).

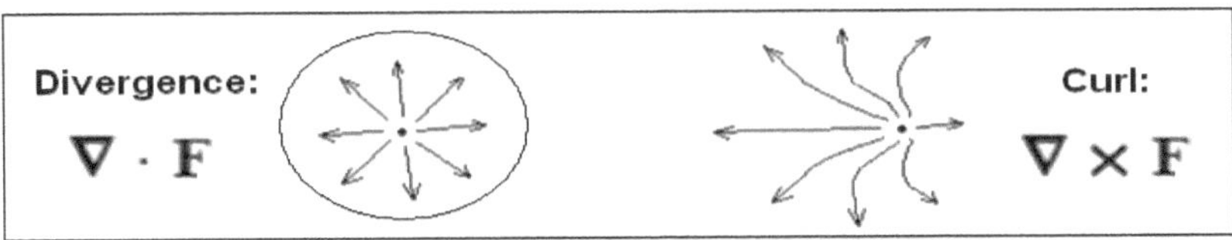

With "q" defined to be our collection of charges, and "ρ" as the charge density (i.e. the amount of "q" per unit of volume), we can relate the *divergence* of our Electric field to its source, "ρ", using what is known as "*Gauss's Law*":

Gauss's Law: If we construct an empty "shell" around a net charge density and measure the total Electric field energy hitting that shell from the net charge, Gauss's Law tells us that the total field divergence/energy we measure hitting that shell will be equal to the total field energy that net charge is radiating (divided by the "scaling factor" of permittivity, with a value that depends on the material it is in, and the system of units you choose to use to describe the field):

Gauss' Law: $$\nabla \cdot \mathbf{E} = \frac{\rho}{\varepsilon}$$

If you think about it, Gauss's Law should make perfect sense in terms of Conservation of Energy, since Gauss's Law tells us that the total amount of Electric field energy measured around the field's source ("ρ"), is equal to the total energy radiated out by that source.

Gauss's Law applied to the Magnetic field looks very similar, except that in this case, the net magnetic charge is zero. The net magnetic charge is zero since there is no such thing as an isolated magnetic monopole charge i.e. all magnets, no matter how small have *both* a North and a South pole. This is true even if we cut a magnet in half in an attempt to isolate one pole, since the resulting two pieces still end up having both a North and a South pole. As a result, you can't have a magnetic monopole analogous to the electric monopole, and therefore Gauss's Law applied to the Magnetic field equals zero:

Gauss' Law:
$$\nabla \cdot \mathbf{B} = \cancel{\mu \rho_B} = 0$$

This brings us *half* way through Maxwell's four equations. To talk about the other two equations (which deal with the curl of one field type as the other type of field changes in time), we need to first introduce another symbol which we will call the "time-change" operator, "$\partial / \partial t$". This operator is very similar to the Del Operator, except that it is used to determine how much something changes over *time* (rather than over *distance*).

So with that we ask, if the Magnetic field changes over time, what does the Electric field do? Faraday's Law tells us that anytime the Magnetic field changes over time, we see a corresponding curl in the Electric field:

Faraday's Law:
$$\nabla \times \mathbf{E} = -\frac{\partial}{\partial t}\mathbf{B} - \cancel{\mu J_B}$$

where "J_B" is the magnetic charge current.

Two things to note about this equation: 1) since we previously established that there are no magnetic monopole charges, the magnetic charge current "J_B" = 0 (we included it here so you can compare it to the next equation). Also, 2) there is a negative sign in front of the "time change operator". Faraday's Law tells us that anytime a Magnetic field changes in time, it creates a curled Electric field with the *opposite polarity* (compared to the magnetic field). This opposing curled Electric field is sometimes referred to as a "counter Electro-Magnetic Force" (EMF), i.e. an induced voltage that *opposes* the creation of the Magnetic field. The negative sign indicates this opposition, and will be demonstrated in our example below.

Finally, we ask what happens to the Magnetic field anytime the Electric field changes in time? Based on symmetry, we suspect a curled Magnetic field to be created. In fact, Ampre's Law tells us that is exactly what happens:

Ampre's Law: $$\nabla \times \mathbf{B} = \frac{1}{c^2}\frac{\partial}{\partial t}\mathbf{E} + \mu \mathbf{J}$$

Where "c" is the speed of light, and "J" is the flow of electric charges (i.e. an electric charge current). Another way of stating Ampre's Law is to say that anytime we have a charge flow (J) *–or–* we have a change in the Electric field over time, *either* event creates a curled Magnetic field.

With all four equations together, we see that the first two equations and the last two equations are almost symmetrical ("mirror images" of each other). The lack of the magnetic monopole is all that denies us that perfect mirrored symmetry:

And God Said...

$$\nabla \cdot \mathbf{E} = \rho$$
$$\nabla \cdot \mathbf{B} = 0$$
$$\nabla \times \mathbf{E} = -\partial_t \mathbf{B}$$
$$\nabla \times \mathbf{B} = \partial_t \mathbf{E} + \mathbf{J}$$

and then there was light.

To demonstrate the kind of insights offered by Maxwell's four *innocent* looking little equations, we offer the following brief examples:

Example 1: power transformers:

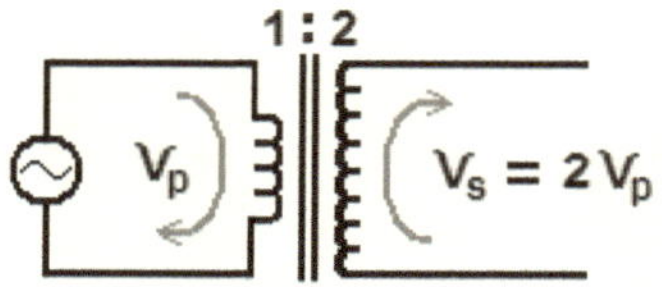

From Ampre's Law, we see that if we have a time-varying current (J) flowing in the primary side of a transformer, we create a curled Magnetic field that consequently is changing in time with the current flow J. As this changing Magnetic field couples into the secondary side of the transformer, Faraday's Law tells us that anytime we create such a time-varying Magnetic field, it creates a counter "EMF" (or an induced voltage) of opposite polarity in the secondary side. If we have twice the amount of wire in the secondary as in the primary (i.e. twice the number of windings in the secondary), we have twice the number of electrons exposed to that time-varying field, creating twice the induced voltage in the secondary (i.e. a step-up transformer).

The negative sign in Faraday's Law tells us that the induced voltage has an opposite polarity and thus opposes the time varying Magnetic field. With a little thought, this too should make sense: If the induced voltage had the *same* polarity, it would *increase* the Magnetic field that created it. This increased Magnetic field would then induce even more voltage, which would then further increase the Magnetic field, etc., leading very quickly to a run-away explosion of energy. Since this is not what we see in practice, but instead see a measurable opposition to current flow, the negative sign is correct.

Example 2: Kirchoff vs. Faraday:

Let us first rewrite Faraday's Law as: (with § meaning "sum around the closed loop").

$$\oint E \bullet dL = -\int \partial_t B \bullet dA$$

Those who have worked with electronic circuits are no doubt familiar with Kirchoff's Voltage Law, which states: "The sum of the voltage drops around a circuit is zero" (note that this is also an expression of the Conservation of Energy, in that it tells us that the voltage dropped by all the loads in a circuit equal the total voltage applied by the battery or generator).

However anyone who has worked near a strong transmitter (e.g. radio broadcast tower, etc.) knows from personal experience that the battery or generator connected to the circuit is *not* the only source of energy entering into that circuit. Truth is, Kirchoff's Law is only a *special case* of Faraday's law, valid only when there is no intruding magnetic field (i.e. **B** = 0):

$$\oint E \bullet dL = -\int \cancel{\partial_t B} \bullet dA = 0$$

Example 3: deriving the Wave Equation:

So now we are in a position to ask "why do we describe the Electric and Magnetic fields propagating as waves"? Because, that is what Maxwell Equations tell us they are:

We start by taking the curl of Faraday's law, and then use a substitution involving what is know as the "BAC – CAB" vector identity:

$$\nabla \times (\nabla \times \mathbf{E}) = -\frac{\partial}{\partial t}\nabla \times \mathbf{B}$$

using identity on Left side:
A x B x C = B A C − C A B

$$\nabla(\nabla \cdot \mathbf{E}) - \nabla^2\mathbf{E} = \frac{-1}{c^2}\frac{\partial^2}{\partial t^2}\mathbf{E} - \mu\frac{\partial}{\partial t}\mathbf{J}$$

$$\frac{1}{c^2}\frac{\partial^2}{\partial t^2}\mathbf{E} - \nabla^2\mathbf{E} = -\nabla\left(\frac{\rho}{\varepsilon}\right) - \mu\frac{\partial}{\partial t}\mathbf{J}$$

In free space there are no charges "ρ", or currents "J", and so the right side therefore goes to zero in free space:

$$\boxed{\frac{1}{c^2}\frac{\partial^2}{\partial t^2}\mathbf{E} - \nabla^2\mathbf{E} = 0}$$

Freespace Wave Eqn.

This result tells us that the rate of change of the Electric field *over time* is equal to the rate of change of that field *across space* (we get a similar result for **B** if we start with Ampre's law). This suggests that **E** and **B** are "undulating" (wave-type) function, such as:

$$\mathbf{E} = \mathbf{E_o}\ \mathbf{Sin(aXt) * Sin(aYt) * Sin(aZt)}$$

where **Eo** is the absolute magnitude of the wave.

If the wave were restricted to propagate solely along a single axis (i.e. aimed in a specific direction), we could describe that wave with something like the following:

$$\mathbf{E} = \mathbf{E_o}\ \mathbf{Sin(aXt)}$$

And that is Maxwell's Equations in a nutshell. Before proceeding to talk about propagation, let's wander into a slightly deeper section of the pool, just to flesh things out a bit:

As we look at these four equations, we almost begin to see what appears to be two paired sets, each composed of two similar and/or related equations: the first two Divergence equations ($\nabla \bullet \mathbf{E}$ and $\nabla \bullet \mathbf{B}$) which are almost identical (with the slight difference explained above), and the two remaining curl equations which clearly have something of a "yen and yang" thing going. We almost can't talk about one field, without discussing the other – their interwoven forms almost suggesting these two fields must somehow be two different "sides" to the same "coin". There is however, the annoying slight difference (e.g. the extra **J** in Ampre's Law) that disturbs this nirvana of symmetries.

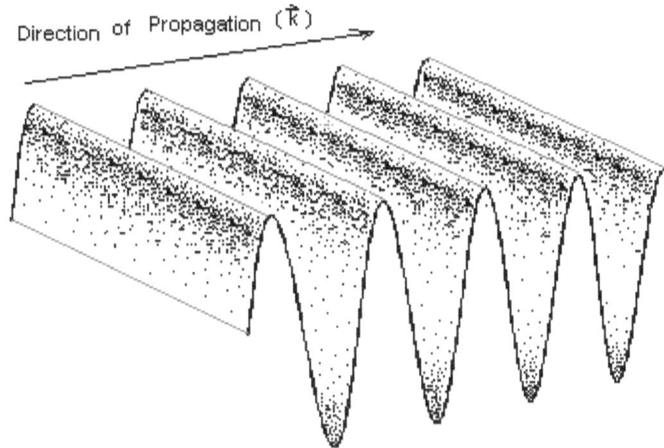

© 2002 "Spread Spectrum CDMA", Steve Lee (McGraw-Hill companies inc.)

Symmetry being as important as it is to Physicists, we would have expected someone would have found a way by now to modify this term to restore our ideal symmetry through some careful mathematical slight of hand or misdirection, if it were possible[30]. However, try as we might, nature stubbornly insists

30 We note that a form of symmetry can be imposed by treating

on denying us the spoils of perfect symmetry, implying some significant physical meaning is tightly bound up in this form. In fact, this was one of Maxwell's greater contribution to E&M, in that he noted that if a capacitor were inserted in our current carrying wire, a charge clearly could not flow past the capacitor. But yet as the charges accumulated on one plate of the capacitor, electrostatic repulsion forced other charges off the second plate of the capacitor and on down the wire. Hence the two terms on the right side of Ampere's law are both viewed as essential, each representing a form of current: **J** as the "normal" current flow, and d**E**/dt being viewed as the "displacement" current "through" the capacitor (which clearly goes to zero if the *change* in the electric field goes to zero).

While on the subject of the quasi-symmetry in Maxwell's equations and the delicate interplay between **E** and **B**, we cite yet another reason why these two must actually be different manifestations of the same electromagnetic "Field-thing". Once again, we point out that if the observer is stationary relative to a charge, that observer only detects an Electric field. However, if a second observer goes flying by this same charge, this moving observer will detect *both* an Electric field *and* a Magnetic field. If there were still other observers that flew past this charge faster than the second along a different course, they would each detect an even different Magnetic field than either of the first two observers.

Obviously there is something wrong with our interpretation of things, since one of the most fundamental rules of the universe is that the laws of physics can *not* be dependent on the observer's frame of reference (otherwise we would end up with an infinite number of laws of physics, which is no better than having none). In other words, the behavior of whatever phenomenon we are observing should *not* depend on how we choose to observe it (e.g. our photographing the moon or a distant nebula should not cause it to become a completely different phenomenon, depending on our act of observation). Reality is what it is, and in the case of the "now you see it, now

---

these two fields as merely being different manifestations of a unified "Electromagnetic" field tensor [4] (see Appendix I).

you don't" Magnetic field, our arbitrary choice of reference frame, mood, planetary alignment, etc. should not cause the phenomenon to be one thing, and then something else.

This inescapable fact, along with the near symmetry of Maxwell's Equations, indicates that **E** and **B** are related, and in fact *intertwined* aspects of the same thing, something we call the "*Electromagnetic Field tensor*". Though we often discuss these two fields separately merely as a matter of convenience (much the same way we break down other complex things to simplify their analysis, e.g. x,y,z components of vectors, etc.), we should understand that they really are only different manifestations of the same Field-tensor.

Another very significant point worth highlighting about Maxwell's Equations, is that they are *linear* differential equations (rather than quadratic "x^2", or cubic "x^3", etc.). This linear behavior implies something very significant about these fields and the way they propagate through various media (e.g. through free space, water, fiber optic cable, etc.): being linear implies they do not combine with other fields within the same medium. Mixing, or "heterodyning" occurs only in a non-linear materials, such as in a diode. This seemingly minor point has *enormous* implications for everything around us, including the fact that all colors of light, all radio and TV channels, all wavelengths of light in a fiber optics cable, etc. do not mix together as they propagate. If they did, absolutely nothing related to light, RF ("Radio Frequency"), or any other electromagnetic energy (including that found in biochemistry and biology) would be the same as it is today!

Another thing worth noting about Maxwell's Equations and what they tell us about electromagnetic fields, is that we often talk about them traveling at the speed of light ("c"). Though it is true that they travel at the speed of light *through empty space*, they do *not* travel that fast when propagating through a region that is anything other than empty space (e.g. through the atmosphere, water, an electronic circuit or transmission line, a fiber optic cable, etc.). What's more, when traveling through non-empty space, they do *not* all travel at the same speed – i.e. *some frequencies/wavelengths travel faster than others*. This difference in behavior is addressed in Maxwell's equations through "ε" and "μ" (the material response factors of "permittivity" and "permeability"). In some cases (e.g. free space), we treat these factors as constants, however in the non-free space case, they are not constants, but instead are typically a function of wavelength. In some cases, they may even be functions of spatial orientation (e.g. birefringence through some plastics, crystals, etc.), in which case they are described using "tensors".

$c = 300 \times 10^6$ m/s speed of light in a vacuum (MKS)
$\varepsilon_o = 8.85 \times 10^{-12}$ C^2/Nm^2 free space permittivity
$\mu_o = 4\pi \times 10^{-7}$ T m/A free space permeability

This "material response" can all be traced back to what we found in Chapter 3 and 4, in that different atoms/molecules respond to different ElectroMagnetic frequencies/wavelengths in different ways due to their Quantum resonances, much the same way a piano/guitar string, flute, or pop bottle resonates at different audio frequencies. This resonance and absorption effect is itself a function of energy stored and energy released (see discussion in Appendix A on resonance).

Bottom line: when an atom or molecule happens to be in the path of a propagating E&M wave that corresponds to some resonant frequency of that atom or molecule, it will absorb some of that E&M wave energy, and then at some later point (after the energy has passed), "disgorge" that energy in a random direction.

As mentioned previously, this non-uniform effect on different frequencies leads to "Dispersion", an effect which causes the various frequencies in a signal to separate one from another, propagating at different speeds and different intensities through a given medium. This has huge implications on the wave energies we transmit, whether that be through an electronic circuit, through the atmosphere, or down a long glass fiber optic waveguide (etc.). Dispersion explains the rainbow/prism effect that takes white light (i.e. light equally composed of all colors/visible frequencies) and separates out each "color" by bending it different amounts as it travels through the glass or raindrop (etc.). We will discuss Dispersion in regards to fiber optics in greater depth in Chapters 7 and 8.

Waveguide and Cavity Modes:

Having derived the E&M wave equation, we are now in a position to talk about E&M waves propagating into confined path (e.g. into a cavity, or down a fiber optic waveguide, etc.). Since a cavity or waveguide represents a confined region, we find that as the energy of a wave propagates from its source point into the cavity or waveguide, it undergoes a reflection at each of the boundaries imposed on the wave by the structure. This reflection of energy tends to set up unique resonant "modes" which are a function of "n" (the index of refraction, i.e. propagation speed), and the size and shape of the cavity (very similar to standing waves on a pulsed rope, or RF transmission line). By using the E&M Wave Equation we just derived, and imposing boundary conditions on that equation which correspond to the shape and dimensions of the cavity or waveguide, we are able to describe the various resonant modes which develop within the structure (see Appendix A for a discussion on resonance). As we shall see in Chapter 7, the resonant modes created within the laser cavity itself have extremely significant impact on both the "quality" of the energy emitted by a laser, as well as how that energy propagates when injected into a "channeling" structure such as a fiber optic waveguide.

In basic terms, we know that when energy is launched into some medium, its propagation speed and behavior is obviously determined by the characteristics of that medium. This doesn't just include its composition, but also its geometric shape, since reflections at its edges have a significant impact on how that wave energy distributes into the media. When the energy moves through any substance (e.g. air, water, glass, etc.), its propagation behavior is a function of the characteristics of that substance (described previously via "permittivity" and "permeability"). When the wave energy reaches any edge in that media – i.e. an area where the propagating characteristics of the media are very different from its previous path – the wave energy finds it very difficult to transition into this new region. Since the Law of Conservation of Energy tells us that this energy has to go *somewhere*, it finds the easiest option is to simply reflect back into the area from which it came. We will show the mathematical description of this phenomenon shortly for completeness, however for the moment we will describe this shape-dependent wave reflection phenomenon, and its *implications*, in a qualitative way in order to get a feel for its impact on our signals.

Since fiber optics cables are cylindrical, let us begin our *qualitative* look by considering the cylindrical shape.

When we impose cylindrical boundaries on our wave, we find that as the energy propagates out towards the wall of the cylinder and reflects, the reflected energy superimposes with the next wave and in the process modifies the *intensity* of the fields *within the structure*. Since the reflections are a function of the cylindrical geometry of the waveguide, the intensity variations within the waveguide take on cylindrically dependent shapes/distributions.

As a result, when we solve the Wave Equation with *cylindrical boundaries* imposed on it, we generate solutions unique to that cylindrical geometry, known as "Bessel functions". These are often described as "Drum head" functions, since they occur anytime a cylindrical drum head is driven into oscillation as energy is injected into it (via a drum strike).

A very effective demonstration of this effect is to take a half-filled round bucket of water and drop a single drop of water or marble into the middle of it. The impact represents energy being injected into the system, which starts waves flowing out from the impact point. If we filmed such an impact in our bucket of water or on a drum head, and then slowed down the images in playback, we would see a circular peak directly in the center slowly rising out of the water or drum as all the reflected waves converge back at the center. Shortly thereafter the rising peak stops, and then begins sinking down past the rest point, where it eventually stops and comes back up to peak again.

This rise and fall effect typically continues several times, gradually diminishing each time as some of the initial energy is slowly lost by various absorption effects. This central peak's shape is due to the fact that we struck the water or drum head at the very center, from which point the impact energy traveled out across the media, struck the walls of the bucket or drum, and was then reflected back to the center. When all the reflections arrive back at the center they *superimpose*, producing one large peak at the center. Mathematically, this effect (wave energy confined to flow within a cylindrical geometry) produces our Bessel function[31], with function zero (J_0) describing the shape of the oscillations on the surface of the water or drum head when struck directly in the center.

If instead of driving the bucket of water (or drumhead) at the very center, we struck it roughly half-way between the center and the wall, we could produce Bessel function J_1. J_1 looks like two counter-poised peaks alternately rising and falling on either side of the center. This effect is again due to the reflections off the drum wall, but since the drum head was struck off-center, the waves traveling to the nearest wall reflect much sooner than the waves traveling to the farthest wall. As a result, they meet on the opposite side of center where their two respective path distances equal. At this off-center point, the two waves superimpose again, creating the same peak and valley effect, only at this off-center location. As they continue traveling, they superimpose once again near the strike point, and in the process create the two counter-poised peaks and valleys on either side of center, forming Bessel function J_1.

The higher order Bessel functions/modes are similarly created in the same way, all again due to the reflecting geometry of the cylindrical media and the movement of the wave energy.

---

31 If we vary the point of entry of the energy injected into such a bounded region, we will generate a whole set of Bessel functions, forming a complete set of "Orthogonal" functions in the process. Other such orthogonal functions can be similarly derived using the wave equation under different boundary geometries, such as "Spherical Harmonics" which are derived using a spherical shaped boundary. These "Spherical Harmonics" are extremely useful in modeling a system with a spherical shape to it, e.g. the atom (see Appendix F).

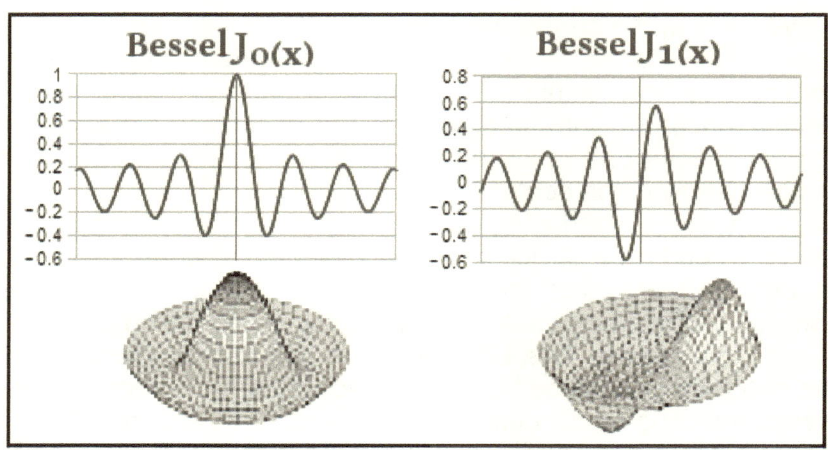

Figure 6.6 Bessel J_0 and J_1.

In the case of a fiber optic cable, the laser used to send light down the fiber is the driving source, and the fiber is the medium. In effect, the cable forms a waveguide, allowing injected energy to propagate down its long axis like a pipe, reflecting off the walls much like our drum head or bucket of water, only distributing the energy through more than just the 2-dimensional surface of the drum/water.

In practical applications, we find that the typical laser does not emit as perfect a beam as we would like (in wavelength, or in modes). The range of wavelengths is due to a number of effects already mentioned related to atomic/molecular Doppler shifting, spontaneous emissions, etc The range of modes in the laser output however is more of a function of the cavity geometry (e.g. see discussion Chapter 5 "Semiconductor" diode lasers). As a result, emissions from many "low-end" diode lasers on the market often lack the ideal wavelength/modal profile. A combination of focusing optics and a very narrow diameter fiber ("single mode fiber") however, can restrict the beam modes to minimize modal dispersion (something that is particularly detrimental in long-haul links; see Chapter 7).

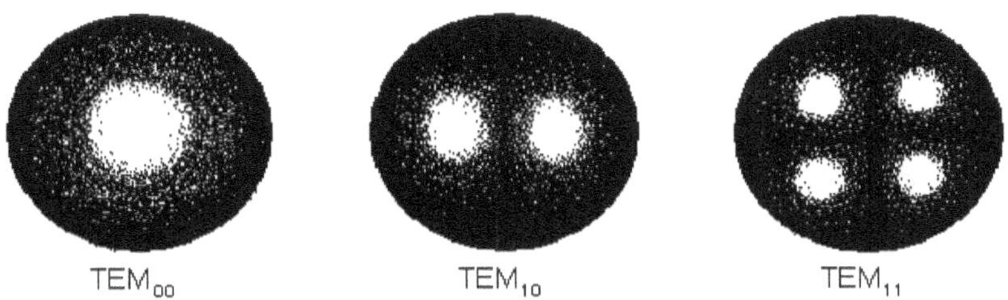

Figure 6.7 cross-sectional view of three possible E&M Modes contained within a cylindrical boundary.

One very important consequence of this modal distribution in a waveguide, is that there is a maximum wavelength (or conversely a minimum frequency) that can be supported by the physical size of the waveguide or cavity. This maximum supported wavelength represents the "*fundamental mode*" of that structure. As with resonance on a string or hollow pipe (etc.), we find that all wavelengths that are longer than this fundamental mode (i.e. that are longer than the physical length of the structure measured in wavelengths), will not couple into the structure very well, and hence die out (as "*evanescent*" waves). This effectively defines a minimum frequency that can be supported for a given structural size, known as the "*cutoff*" frequency. All frequencies below this cutoff frequency, cannot be supported by the waveguide or cavity, and hence soon die out if injected into the structure (see Chapter 7).

To begin our *mathematical* description of these resonant modes within a cavity, we first start by defining a cavity shape that will be easy to work with using a Cartesian coordinate system (for simplicity). To that end, we choose a rectangular box, with length "L", width "W" and height "H". For simplicity, we will assume that there are no sources within the cavity. Thus our "zero source" wave equation is:

$$d^2\,\mathbf{E}/dx^2 + d^2\,\mathbf{E}/dy^2 + d^2\,\mathbf{E}/dz^2 - 1/c^2\; d^2\mathbf{E}/dt^2 = 0$$

If we assume a harmonic time dependence of "exp(-iωt)" (see Appendix D) and write our field as separate "x", "y" and "z" components:

$$\mathbf{E}_{(x,y,z,t)} = E_{(x)}\ E_{(y)}\ E_{(z)}\ \exp(-i\omega t)$$

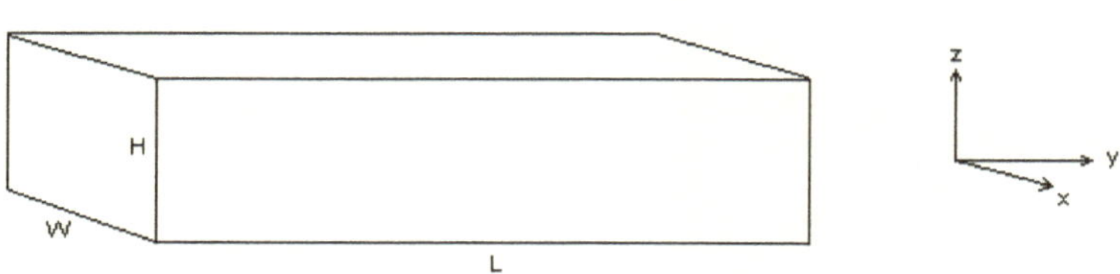

Figure 6.8 Rectangular cavity used in our waveguide example.

We now substitute this form into our wave equation. When we do, our wave equation becomes:

$$[\ E_{(y)} E_{(z)}\ d^2 E_{(x)}/dx^2 \quad + \quad E_{(x)} E_{(z)}\ d^2 E_{(y)}/dy^2$$

$$+\ E_{(x)} E_{(y)}\ d^2 E_{(z)}/dz^2\ ] * \exp(-i\omega t) - (1/c^2)\ d^2\mathbf{E}/dt^2 = 0$$

Defining a "wave number" "k" = ω/c = 2π/λ and performing the differentiation in just the last term, we have:

$$(1/c^2)\ (-i\omega)^2 \exp(-i\omega t) = -k^2 \exp(-i\omega t)$$

Substituting this back into our wave equation and dividing through by "$E_{(x)}\ E_{(y)}\ E_{(z)}$* exp(iωt)", our equation reduces to:

$$(1/E_{(x)})\ d^2 E_{(x)}/dx^2 + (1/E_{(y)})\ d^2 E_{(y)}/dy^2$$

$$+ (1/E_{(z)})\ d^2 E_{(z)}/dz^2 + k^2 = 0$$

We note that each term is now expressed as a function of only one variable; if we assume k^2 can also be separated into components:

$$k^2 = k^2_x + k^2_y + k^2_z$$

This allows us to now separate and solve each term independently:

$$(1 / E_{(x)})\ d^2 E_{(x)}/dx^2 = k^2_x$$

$$(1 / E_{(y)})\ d^2 E_{(y)}/dy^2 = k^2_y$$

$$(1 / E_{(z)})\ d^2 E_{(z)}/dz^2 = k^2_z$$

Using Euler's relation (see Appendix D), we can write a solution for the wave in the form of "e^{ikx}":

$$E_{(x,t)} = E_{0\,(x)} \exp(ik_x x) \exp(-i\omega t)$$

$$E_{(y,t)} = E_{0\,(y)} \exp(ik_y y) \exp(-i\omega t)$$

$$E_{(z,t)} = E_{0\,(z)} \exp(ik_z z) \exp(-i\omega t)$$

Here we apply the boundary condition of the rectangular waveguide, which requires that the Electric field goes to zero at the boundary (e.g. $E_{(x,t)} = 0$ when $x = 0$ and L). This implies that: $k_x L = n_x \pi$ (where n_x is an integer which ranges from 0 to infinity), i.e. $k_x = n_x \pi / L$. As a result, our solutions become:

$$E_{(x,t)} = E_{0\,(x)} \exp(i\ x\ n_x \pi / L)\ \exp(-i\omega t)$$

$$E_{(y,t)} = E_{0\,(y)} \exp(i\ y\ n_y \pi / W)\ \exp(-i\omega t)$$

$$E_{(z,t)} = E_{0\,(z)} \exp(i\ z\ n_z \pi / H)\ \exp(-i\omega t)$$

Since $k = \omega / c$ = SQRT [$k^2_x + k^2_y + k^2_z$], and since "$\omega = 2\pi F$", then: $F = kc/2\pi$, therefore:

$$F_{x,y,z} = c / (2\pi) \text{ SQRT } [k^2_x + k^2_y + k^2_z]$$

$$= c / 2\pi \text{ SQRT } [(n_x\pi / L)^2 + (n_y\pi / W)^2 + (n_z\pi / H)^2]$$

Example:

Find the frequency for the fundamental mode "1,0,0" of the resonant cavity, if "L" = 9 cm, "W"= 10 cm, and "H" = 11 cm.

$$F_{x,y,z} = c / (2\pi) \text{ SQRT } [(n_x\pi / L)^2 + (n_y\pi / W)^2 + (n_z\pi / H)^2]$$

$$= c / (2\pi) \text{ SQRT } [(\pi / 0.09)^2 + 0 + 0]$$

$$= 1.667 \times 10^9 \text{ Hz}$$

Example:

Using the same cavity, find the frequency for the "0,1,0" cavity resonant mode.

$$F_{x,y,z} = c / (2\pi) \text{ SQRT } [0 + (\pi / 0.10)^2 + 0]$$

$$= 1.50 \times 10^9 \text{ Hz}$$

Example:

Using the same cavity, find the frequency for the "2,0,1" cavity resonant mode.

$$F_{x,y,z} = c / (2\pi) \text{ SQRT } [(2\pi / 0.09)^2 + 0 + (\pi / 0.11)^2]$$

$$= 3.6015 \times 10^9 \text{ Hz}$$

Using a similar approach, we can generate solutions for a cylindrical cavity. With that geometry however, we find that using the Cylindrical coordinate system (r,θ,z) is much more useful than the Cartesian (x,y,z) coordinate system. In any case, when we impose the boundary condition that the Electric field goes to zero at "r = R" (at the edge of the cylinder wall), we generate solutions unique for that geometry, i.e. Bessel Functions (see Arfken [3]). As we saw in our rectangular waveguide examples above, the solutions for a cylindrical waveguide also produce a discrete range of independent resonant modes in "r", "θ" and "z" analogous to the "x,y,z" modes demonstrated in the above examples.

Wave propagation revisited:

Returning to our E&M wave equation: using Euler's relation again (see Appendix D), we can write a solution for the *ideal monochromatic* wave in the form of "e^{ikx}":

$$E_{(x,\,t)} \;=\; E_{(k)}\; e^{ikx - I\omega t}$$

Using k = ω/v = 2π/λ (where "v" is velocity), this becomes:

$$E_{(x,\,t)} \;=\; E_{(k)}\; e^{ik(x - vt)}$$

In the non-ideal (*real world*) case, all waves generated by *any* known sources are *never* truly monochromatic, but instead have a *range* of frequencies/wavelengths. That being the case, we need to write our generated field as a composite *sum* of tiny "slices" of frequency space "dk", and then "sum up" all of those "dk" contributions (very analogous to what Planck did with Black Body radiation), giving us the following form:

$$E_{(x,\,t)} = \int E_{(k)}\; e^{ik(x - vt)}\; dk$$

Those familiar with Fourier techniques (see Appendix B) may recognize this has the precise form of a "Fourier Transform" – a mathematical wave "tool" that allows us to "toggle" our description of the field "E" back and forth between a view of its physical distribution over space and time, verses a view of this same field as a distribution of frequencies (in terms of "k"), i.e. its *spectral distribution*:

$$E_{(k)} = \int E_{(x,t)}\; e^{i k (x - v t)}\; dx$$

Adopting such Fourier techniques for analyzing "wave-like" things, we can begin to view the propagation of our wave as a sequence of "transforms" produced by the media the wave moves through (rather than merely seeing the equations related to propagation as nothing more than ugly raw math). Such use of transforms has become a common technique in Systems Engineering, in which an overall system process is simply redrawn as a "black box" that performs some well-known operation on an input signal, to produce a refined output (not unlike our Carnot engine).

In the parlance of Fourier Mechanics, we can often view such "processing" of our wave energy as an example of a "Fourier Transform" or even a "Fourier Convolution", allowing us to describe the overall effect on our signal as it propagates through some well-defined "black box" processor, be that a coaxial cable, a radio transceiver, the atmosphere, a lens or aperture, or a fiber optic cable, (etc.).

The advantage that such an approach offers, is that it brings to bear the entire arsenal of tools and insights developed by the analytic techniques of Fourier Wave Mechanics to provide a wider understanding into the behavior of the system under study – via a temporal transform (time space), a spectral transform (frequency space), and even as a "convolving" transform on the propagating wave by some element in its path (e.g. a lens effect, dispersion by resonance/absorption, diffraction through an aperture, etc.).

In a convolution transform, the *distribution of energy* that is our signal is treated as being *convolved* with the transforming function of some well-defined "black box" "transformer", producing a "daughter" product that inherits characteristics directly related to the two "parents" (i.e. the original E&M wave and the convolving lens, aperture, gas, etc.).

If for example the medium's transforming function includes a well-known frequency-dependent absorption (e.g. due to molecular resonances, etc.), or frequency-dependent lensing or diffraction effect, (etc.), this known behavior can be established as part of the effect of this "black box" transformer, and simply built into our model, rather than our having to go through the gory integration process every time we need to evaluate the propagation.

Of course a few diehards might object that the transform approach appears somehow less rigorous than grinding out the integrals. However we would argue that it is no less rigorous than using integrals built around approximations such as the "Fresnel" or "Fraunhofer" simplifications (see below). In addition we find that Fourier analysis offers a certain amount of clarity not obtainable otherwise, and this allows one to "see the forest, in spite of the trees", so to speak.

At the other extreme, some students tend to be somewhat obsessed with the need to convert all integrals or differentials they encounter into a number (e.g. "42", for those so obsessed), insisting that this somehow brings "closure" to the process and allows them to sleep at night. To these individuals we argue that the symbolic form embodied in Fourier Mechanics provides a great deal more meaningful information as a whole than can be obtained by any single extruded number. Unfortunately, many textbooks on the subject often tend to reinforce this number fixation by tending to show only examples where all the integrals reduce down to something very tidy and "complete". In reality, most such presentations tend to be rather artificial (for the sake of demonstration), since most problems in the real world are not so easily reduced to such a simplistic "mono-syllabic" grunt of an expression.

By weaning oneself away from the obsessive-compulsive tick that insists on grinding down the integral expression until it extrudes some hard and fast quantifiable result, and accepting the possibility that the symbolic representation could offer a profoundly meaningful statement about the global behavior of the system, a "bigger picture" view of things can slowly begin to emerge. Such a global perspective conveys a great deal more information than the mere particle of fact that a single number or even algebraic expression might provide. Fourier Mechanics offers such a perspective, and is well worth the effort to pursue (see Appendix B).

Diffraction:

The term "diffraction" effectively implies the *bending* of waves around an obstacle or through some aperture. A good demonstration of this phenomenon occurs when monochromatic light is shot through a pair of apertures, as in Young's Double Slit experiment, forming a diffraction pattern on a surface opposite the apertures (see Figure 3.7). This diffraction effect can also be demonstrated by dropping a pebble in a pond and watching the ripples it generates as they encounter any solid objects in their path (e.g. a rock, or pole, etc.). The impact of the pebble on the water imparts energy into the pond, which then travels through the water in the form of waves. As these waves encounter any solid objects in their path that are at least comparable to the wavelength of the ripples, some portion of the energy in the wavefront is reflected by the object (since the object has significantly different "propagation characteristics" compared to the water, the wave energy cannot couple into it, and so it reflects). The remaining energy in the wavefront will flow *around* the object and continue forward at the same speed and in roughly the same direction as it was moving prior to encountering the object (see Figure 6.9).

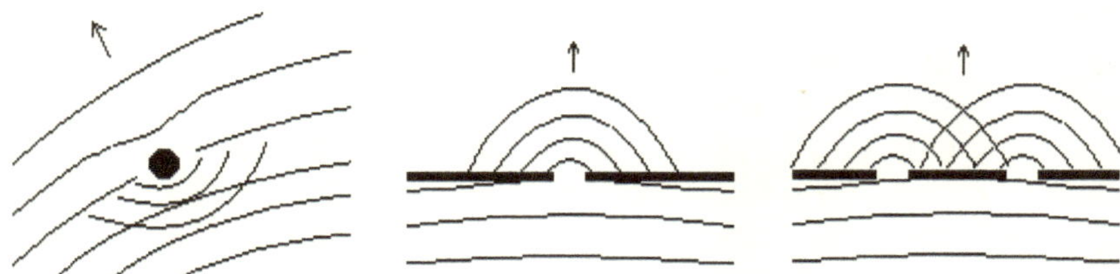

Figure 6.9 Huygen's Diffraction model of wave propagation.

If instead of a simple rock or pole we placed a large wall with an opening at its center (again comparable in size to the wavelength of the ripples), we find that when the ripples reach the wall, only the area of the wavefronts that aligns with the opening makes it through the aperture and begin to radiate into the other side. Once through this aperture, the energy that "coupled" through then begins to radiate out from that point as if the aperture were an original source. (Note that if the opening is much smaller than the wavelength, very little wave energy couples through the opening.)

Both of these behaviors are examples of diffraction, and have direct counterparts to the propagation of E&M waves through or around objects (e.g. through the metal cross beams in a bridge, over the top "knife edge" of a mountain, or through a window of a building, around trees or billboards, etc.). To explore this particular aspect of propagation, we borrow a technique from optics known as Huygens' principle, in which we consider each point on the surface of our wavefront as if it acts as a tiny source (again analogous to Planck's approach to the Black Body problem). These tiny sources then superimpose to produce the next wavefront across a new surface area "da".

We begin by restating the solution to the E&M wave equation using "$e^{i\,\mathbf{k}\,\mathbf{x}}$" as done previously, considering only the magnitude of the field, not the vector direction. Next, we omit the time term "exp(-iwt)", with the understanding that variations over time are implied, so that we might concentrate on the spatial distribution as the wave propagates through the medium:

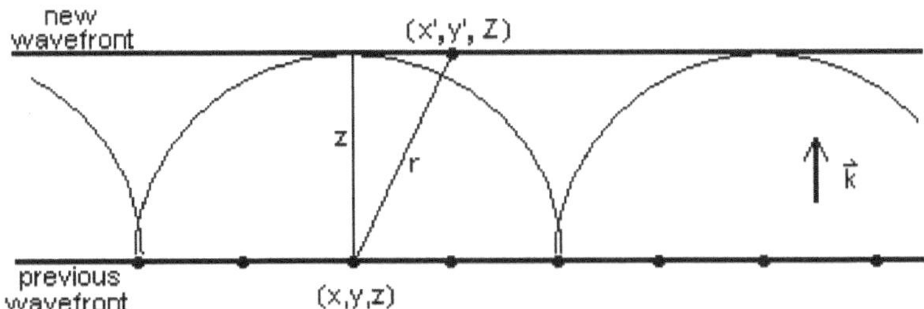

Figure 6.10 Huygen's wavefront propagation model.

$$E(\mathbf{r}) \quad = \quad E_o \ \exp(i\ \mathbf{k}\bullet\mathbf{r})$$

(where: $k = \omega/v = 2\pi/\lambda$, and E_o is the magnitude of the field.)

In Huygens' model we assume each tiny point of the wavefront generates small spherical wavelets, each with a fraction of the overall field strength "dE(**r**)". The field element itself has some inverse relation to wavelength and distance, on the order of $1/\lambda r$. The sum total effect of all these small wavelets create the next wavefront across the surface "da" of the new wavefront:

$$dE(\mathbf{r}) \approx \mathbf{E_o} \; 1/\lambda r \; \exp(i\,\mathbf{k \bullet r})\; da$$

If we draw our small wavelet spheres with a radius of "Z", we can represent a position on the combined wavefront "r" in terms of x, y and Z:

$$r = (x^2 + y^2 + Z^2)^{1/2} = Z\,[1 + (x^2 + y^2)/\,Z^2\,]^{1/2}$$

If we consider only small "x" and "y" distances compared to the radius of each wavelet (Z), we can use a Taylor approximation[32] to rewrite this into:

$$r \approx Z + (x^2 + y^2)/\,2Z$$

therefore "k" (where $k=\omega/c$) times "r" is:

$$kr \approx kZ + k(x^2 + y^2)/\,2Z$$

We now rewrite our new wavefront field as:

---

32 A Taylor series approximation uses the same "summation of component" technique used in a Fourier Superposition to represent a function F(x=a) as a sum of distinct components:

$$F(x) = F(a) + (x - a)\,F(a)'/1 + (x - a)^2\,F(a)''\,/\,2! + \ldots$$

where the primes indicate 1st, 2nd, 3rd ... derivatives at x=a, and "n!" implies "n Factorial" (e.g. 5! = 1*2*3*4*5). Hand held calculators typically use such representations to compute trig functions, etc. rather than store an infinite number of values in a table for each function.

$$dE(\mathbf{r}) \approx \mathbf{E_o} / \lambda z \; * \exp(ikZ) \exp(ik \, [(x-x')^2 + (y-y')^2] / 2Z) \; da$$

If we want to know what the field E(**r**) is at a distance "r" away from the source, we must integrate the above expression (we let "da" = dx dy), which gives us the following form:

$$E_{(r)} \approx \exp(ikZ)/\lambda z \; {}_{-\infty}\!\int\!\int^{\infty} \mathbf{E_o} \exp(ik \, [(x-x')^2 + (y-y')^2] /2Z) \, dx' \, dy'$$

This is the *Fresnel Diffraction Approximation* (pronounced "Fre' nel") and approximates our field rather well at distances where $r << \lambda$ (i.e. $kr << 1$; a region referred to as the "Fresnel zone", or the "Near zone").

When distances are much greater than this, the assumption we made for our Taylor approximation ceases to be valid, and so too then does the Fresnel approximation. For distances where "r" is very large compared to wavelength, we switch to an alternate approximation known as the *Fraunhofer diffraction approximation,* which is valid only when:

$$Z >> k\,(x^2 + y^2)$$

i.e. when $k\,(x^2 + y^2)$, which is a function of wavelength, is very small compared to distance (i.e. $kr >> \lambda$) – a region referred to as the "Fraunhofer zone", or "Far Field zone" (which by rule of thumb is when "r" is roughly $10*\lambda$ or better). Using this approximation, the term:

$$\exp(ik \, [x^2 + y^2] / 2Z) \approx 1$$

With this approximation, our field equation becomes:

$$E(\mathbf{r}) \approx \exp[(ikZ) + (x^2 + y^2) / 2Z] / \lambda z$$

$$* \; {}_{-\infty}\!\int\!\int^{\infty} \mathbf{E_o} \exp(-ik \, [xx' + yy'] /Z) \, dx' \, dy'$$

This Fraunhofer approximation represents our field in the "Far Zone". We note that we retained our "(xx'+yy')/Z" term in the exponential despite being small, since when dealing with refraction, it is the *fraction* of a wavelength (not the total distance) that leads to the constructive and destructive interference of the wave that produces the diffraction pattern.

If we step back and look at the diffraction integral prior to any approximations for a moment, we note again that it has the form of a Fourier Transform:

$$E(\mathbf{r}) = \int E_{(\mathbf{r}')} \exp(-i\,\mathbf{k} \bullet \mathbf{r}')\, dr'$$

only in this case, we are not transforming between "time" and "frequency", but rather "r" space and "r' " space (i.e. on either side of the diffracting object). The diffraction effect comes as a result of the constructive and destructive interference as "**k** • **r**' " moves through fractions of a wavelength. We find the Fourier transform treatment of this aspect of propagation again simplifies the analysis, by allowing us to view the diffraction effect as something of a "black box" transformer (rather than just another ugly integral expression).

Chapter 6 Review questions:

1. What prompted our need to introduce the notion of "fields"?
2. What is a vector? Why do we separate a vector's components? Why is this valid?
3. What does Coulomb's Law tell us?
4. What is the Lorentz force? What does it tell us about the relationship between the Electric field and the Magnetic field?
5. If we have a collection of charges, will it always produce an Electric field? If those charges are "not moving", is there a Magnetic field? (trick question).
6. Can you have an Electric field without a Magnetic field (trick question).

7. Explain each of Maxwell's equations. What do they tell us about the relationship between the Electric field and the Magnetic field?
8. Explain the implications of the fact that Maxwell's equations are linear.
9. Use Maxwell's equations to explain an AC transformer.
10. What is the classical E&M Wave Equation and how is it related to Maxwell's equations?
11. What is a waveguide, and what effect do boundaries (walls) have on the propagation of energy down such a waveguide.
12. How does the wave equation help describe the energy injected into a fiber optic cable?
13. What are "Evanescent" waves and how do they relate to waveguides?
14. What is "diffraction"? What causes it?
15. What is dispersion? What affect does it have on pulses of energy?
16. Explain the Taylor Expansion Approximation in terms of "a sum of components" (i.e. superposition).
17. What is the "Far Field", and what approximation did we make for this field in terms of wavelength? Why is this valid?
18. What is a Fourier Transform, and how does treating the propagation and diffraction integrals as "black box" transformers help simplify the modeling of propagation through a given medium (e.g. through a fiber optic cable, over a "knife edge", through an aperture, etc.)?
19. Would the rectangular cavity described in the chapter examples support a 1.2×10^9 Hz wave? Explain.
20. What length would we need in this cavity to support a 1.0×10^9 Hz wave in the "1,0,0" mode?
21. What would happen if we injected a 1.8×10^9 Hz signal into the cavity in the previous question? A 2.0×10^9 Hz signal?
22. Prove that "$\mathbf{E}(x,y,z,t) = E_0 \, Sin(xt)$" is a valid solution to the free space wave equation (where "E_0" is the wave's maximum amplitude).
23. Prove that "$\mathbf{E}(x,y,z,t) = E_0 \, e^{-i x t} \, e^{-i \omega t}$" is also a valid solution to the free space wave equation.

Chapter 7 Physical Components

In our first chapter, we discovered that to create lasing, we needed little more than a productive gain media (one with a well-defined metastable state at the desired output wavelength), and an effective pumping mechanism[33]. As it turns out, just about anything that meets these two simple requirements can be made to lase (e.g. plastics or liquids saturated with fluorescent dyes, pockets of Carbon Dioxide (CO_2), or Ammonia (NH_3), or Nitrogen (N_2) gas cells, etc). Alone however, most of these produce only weak outputs ($\sim 10^{-12}$ W or less). The challenge then, is in producing a useful output power density at a reasonable cost (i.e. a system that lases in a reasonably efficient way). What we need is a more *effective* way to amplify the photons that are released from the gain media, be that HeNe, CO_2, NH_3, N_2, etc.

To make a more effective laser, we need to excite as large a number of gain media atoms / molecules as possible, in close enough proximity as possible to have a reasonably good chance of stimulating each other into a coordinated release of photons. If we make the gain media too sparse, we will decrease the probability that a released photon will encounter another excited atom/molecule before the neighbor atom/molecule spontaneously de-excites, resulting in a weaker output than if we used a more dense gain medium.

To increase the probability that a released photon will encounter multiple excited atoms, we could place this concentrated gain media in a *very* long tube that is fully exposed to the pumping energy throughout the length of the tube, and then hope some percentage of our photons will travel down the long axis, stimulating other atoms in the process. At the end of the cavity we would then simply need a device to channel the

---

33 In the case of the NASA Martian findings cited in the introduction, the lasing media was Carbon Dioxide known to be present in the Martian atmosphere, while the pumping mechanism was simple exposure to energy radiating from the sun [2].

released energy as needed (e.g. lens, mirror, etc.). Unfortunately, the fact that the cross section of a photon is extremely small (smaller than the diameter of an atom), implies this long tube approach would not generate more than perhaps a few million or billion photons per centimeter, i.e. an output in the pico-Joule realm or less (e.g. 10^9 photons * h * 10^{14} Hz = 60×10^{-12} Joules).

However, if we could somehow cause the photons to make *multiple passes* through the excited gain media, we could very easily increase the number of photon-atom interactions by several orders of magnitude with very little effort and thereby increase the output power of the laser considerably. This approach would also allow us to shorten our tube dramatically from the "long" tube scenario, since we no longer need length to provide the necessary atom-photon interaction.

In practice, this is exactly how most laser systems operate, using a relatively short vessel of gain media with reflectors of some type (e.g. simple mirrors) at either end of the vessel. To work correctly of course, we have to make sure that the mirrors are well aligned such that they reflect the photons back into the gain media rather than out of it. But considering the potential for much higher output powers, aligning the mirrors seems like a very small price to pay.

The caveat here is that by closing two ends of the active region, we effectively create a "resonant cavity", and therefore we need to consider the effect on the laser system that such a *resonance* will impose on it (see Appendix A). All cavities (e.g. organ pipes, pop bottles, or laser cavity, etc.), resonate anytime their length "L" exactly matches a half wavelength of the energy injected into them *-or-* some integer multiple of a half wavelength:

$$L = n(\lambda / 2) \qquad n = 1, 2, 3, \ldots$$

where $\lambda = v / \text{Freq}$, and v = the velocity of propagation of the wave inside the cavity (e.g. for sound, $v \sim 345$m/s, for E&M waves, "v" is approximately equal to "c" = 300×10^6 m/s for a very thin gas, and decreases as the gas density increases).

For example, if we had an organ pipe that resonated at 100 Hz, λ = 345/100 = 3.45 meters (therefore λ/2 = 1.725), it will also resonate at 200 Hz (since its wavelength is half as long), and at 300 Hz (since its wavelength is a third as long), etc. As mentioned in Chapter 6, there is a minimum frequency at which a cavity will resonate, which corresponds to n = 1 in our above equation, since any frequency lower than this will equate to too long a wavelength to physically fit the length of the cavity. This minimum frequency is known as the "*Cutoff Frequency*" for the cavity.

Though the cavity has a minimum supportable frequency related to its length, its physical dimensions do not tend to impose the upper end on the range of supportable frequencies. As a result, "n" could *theoretically* range to infinity. In practice however, a number of other factors tend to limit the range of supportable frequencies in the cavity, and hence these higher frequencies (referred to as "*Harmonics*") tend to diminish as "n" increases.

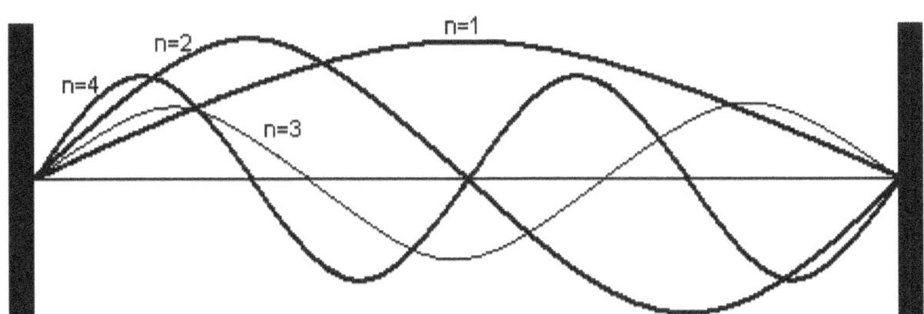

Figure 7.1 Several resonances in a given physical length.

Since the typical wavelength for many types of lasers measures in the nanometer (10^{-9} m) range or smaller, we know that typical cavity dimensions correspond to a great many wavelengths (for example, the HeNe laser example cited in Chapter 5 had an output wavelength of 632.8 nm, and a cavity length of 10 cm, which corresponds roughly to: 0.1m / 633 nm =

160,000 wavelengths). As a result, such a very large cavity (in terms of wavelengths) tends to support a very large number of resonant modes.

The separation between modes in a laser cavity can be calculated by re-writing the previous relation between length "L" and wavelength as: $\lambda = n\,(L/2)$ and using: $Freq = c / \lambda$. Therefore we find:

$$Freq = c\,n / (2\,L)$$

Therefore, the separation in frequency between two successive cavity modes is:

$$\Delta freq = c\,n / (2\,L) - c\,(n+1) / (2\,L)$$

$$= c / (2\,L)$$

Example: 10 cm cavity laser

What is the separation in frequency between two cavity modes if the cavity length L = 10 cm?

$$\Delta freq = 300 \times 10^6\ M/s\ / (2 \times 0.1M) = 1500\ MHz$$

In Chapter 5, we briefly described the pumping process of the gain media in terms of a simple "balance sheet" operation. In that discussion we found that the amount of energy entering the cavity had to at least equal all the energy leaving the cavity, as well as compensate for the internal losses. In practice we find that this energy exchange is very wavelength dependent.

As a result of this wavelength dependence, some wavelengths in the cavity receive sufficient energy to exceed the minimum pumping threshold for steady state operation, while other wavelengths do not, and as a consequence die out. This effect helps define the upper limit on the "viable" wavelengths that can be sustained in steady state within the laser system.

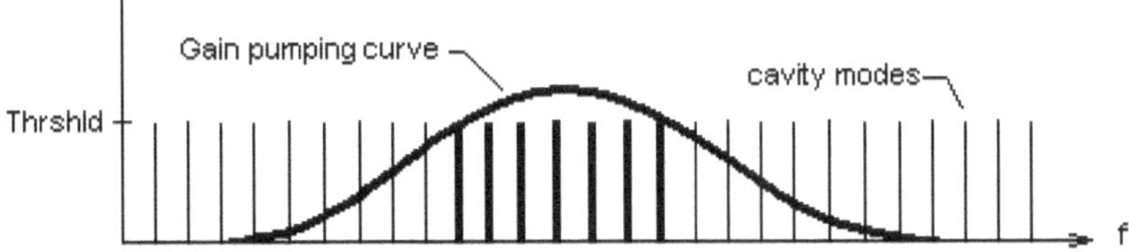

Figure 7.2 Plot of frequencies that meet or exceed the minimum pumping threshold.

Example:

Continuing from the previous example, assume the gain pumping threshold spectral width is 1nm (= 300x10^{15} Hz). How many viable modes are possible in the cavity used in the previous example?

Number of modes = line width / (frequency separation between modes)

= 300x10^{15} Hz / 1500 mHz

= 200 x 10^{6} viable modes (!)

In many applications, a laser beam composed of such a large number of "spectral peaks" is very undesirable (e.g. long-haul fiber optic links require a pure, single wavelength light source for optimum performance; see discussion in Chapter 8). Since the resonant effects of the cavity itself can significantly impact the "wavelength purity" of the laser emissions, the design of the laser cavity needs to be carefully considered. Therefore, we will now present something of an overview of the more prominent components in the laser cavity that affect both the spectral purity and the modal content.

Mirror:

The first cavity component we will describe is the mirror. So far in our discussions, we have encountered several different techniques which have been developed which directly affect the efficiency of a laser's operation, including using "impurity" gases (such as Helium in the HeNe laser, and Nitrogen in the CO_2 laser, etc.), to help channel energy into the desired metastable state.

In our description of the semiconductor diode laser, we found that one of the simplest things we can do that has a very significant impact on the overall effectiveness and efficiency of the laser, is to insure that the mirrors we use to define the cavity of the laser are highly reflective. When these mirrors are absent or only partially successful at reflecting photons back into the cavity, the ratio of power out of the device vs. the amount of raw energy we have to inject into it is very low. In other words, when the mirrors fail to redirect photons back into the middle of the excited gain media, the efficiency of the laser suffers dramatically.

This requirement may seem a little odd at first glance, since one might think that getting photons out of the laser is better than throwing them back into it. But remember, the core aspect at the heart of the laser is that it is a photon *amplifier*; the more photons we have knocking around inside the excited gain media, the more stimulated emissions they will generate (i.e. for every 1 photon we throw back into the gain media, we get 10 or 50 or 100 photons out), up to the point where we begin to deplete the population inversion. To make optimum use of the raw energy we invested into exciting the gain media, we must stimulate a coordinated release of energy from as many excited atoms (or molecules) in the gain media as possible. Otherwise, the energy that we worked so hard to store in the metastable state will not contribute to the coherent beam energy. Though it will eventually be released out of the excited media after the "lifetime" of the metastable state, these emissions will typically not be in phase with our output beam, and hence tend to be counterproductive. Consequently, we need to enclose the gain media within at least two well-aimed mirrors to ensure we extract as much of the energy we pumped in as possible.

Of course when we do place mirrors around the gain media, we can't just throw them into the system with complete abandon, since *what* we use for mirrors, and *how* we use them will dramatically affect how successful we are at redirecting photons back into the midst of the inverted gain media. If for example, we use two plane mirrors which are not perfectly parallel, all released photons will eventually be reflected out of the cavity after only a few passes between the two mirrors.

To optimize the reflection process, we need to either be absolutely certain the two plane mirrors are perfectly parallel (throughout the operating temperature range of the laser no less), or use mirrors that are slightly concave, such that any photons that are traveling slightly off the central axis of the cavity, are reflected back inward towards the center of the cavity (see Figure 7.3).

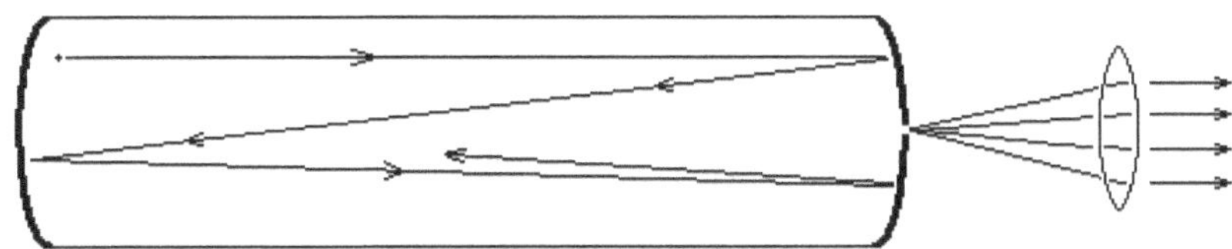

Figure 7.3 Cavity using concave mirrors.

Matrix design approach:

In order to simplify the design process we will use a first approximation description of the E&M wave through the cavity, treating this energy flow as a simple straight line "Ray Optics" path, while neglecting all the fine detail aspects of the propagation (including diffraction effects, etc.). This description is known as "Geometric Optics", and though it is only a simplified *approximation* which neglects the more subtle wave-related aspects of E&M propagation, it will be more than adequate for our cavity design purposes.

We will now write the various components in our laser cavity (e.g. light rays, lens, free space propagation, mirrors, etc.) as either a vector (in the case of the light ray), or as some "black box" transforming operator (written as a matrix) which acts on the ray vector as it propagates. In a real sense, this approximation is very analogous to the method mentioned at the end of Chapter 6, in which we stated we could describe any wave propagating through a region or material, as undergoing some "black box" transformation by the propagating media. In effect, we are now going to put a simplified definition to that "black box" transformer. As the energy in the E&M "ray" enters one of these transformers (for example a lens or gas cell, etc.), the transformer performs some well-defined operation on it, modifying ("transforming") the ray in a well-known and thus very predictable way.

This type of analysis allows us to gloss over the questions of "why" or "how" it does what it does, and go straight to the details of describing the end effect of the transform process, and thereby provide us a simple and yet concise method of modeling the overall progress of the ray through the system. One of the greatest advantages this approach offers, is that it eliminates all the mess of performing any ugly differentiations or integrals. And since matrix operations are little more than a sequence of multiplications in a well-defined sequence and order (see Appendix I), the process can be easily automated via a computer program[34].

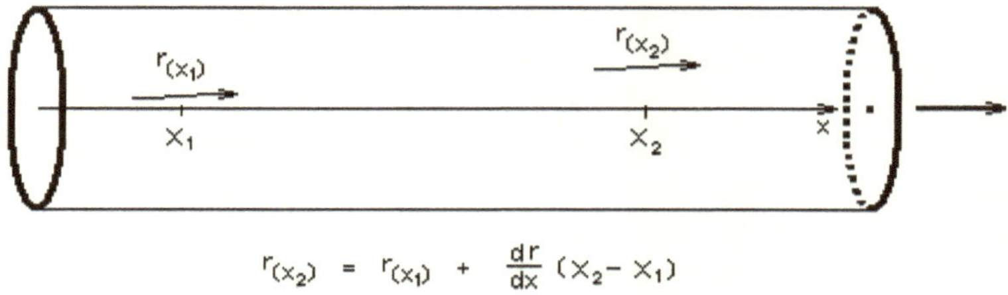

Figure 7.4 Ray vector in a cavity at two positions (where dr/dx = slope).

---

34 There are a number of good references available which describe this matrix approach in greater detail, including Gerrard and Burch [10] as well as Millonni and Eberly [6].

In essence, this matrix approach effectively compacts the standard equations that we would use if we did this "long hand". For example, a ray going from one end of the cavity (point "1" in Figure 7.4) to the other (point "2"), travels distance "D" through a slope "dr/dx", following the equation of a straight line (y = b + mx, where "m" equates to slope):

$$r_2 \quad = \quad r_1 \; + \; dr/dx \;\; D$$

To begin, let us first construct a "black box" matrix for each element, including the ray vector, composed of the position component "r", and the slope (or rate of change) of the position vector "dr/dx", as shown in Figure 7.4.

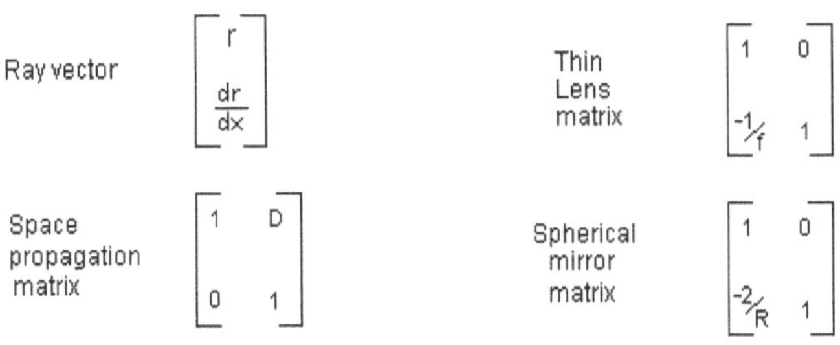

Figure 7.5 matrix elements for ray tracing, where "D" is distance, "f" is focal length, and "R" is radius of curvature.

Figure 7.6 Possible cavity geometries for example problems.

Space propagation
$$\begin{bmatrix} r \\ \frac{dr}{dx} \end{bmatrix} = \begin{bmatrix} 1 & D \\ 0 & 1 \end{bmatrix} \begin{bmatrix} r \\ \frac{dr}{dx} \end{bmatrix} = \begin{bmatrix} r + D\frac{dr}{dx} \\ \frac{dr}{dx} \end{bmatrix}$$

Space + Thin Lens
$$\begin{bmatrix} 1 & 0 \\ -1/f & 1 \end{bmatrix} \begin{bmatrix} r + D\frac{dr}{dx} \\ \frac{dr}{dx} \end{bmatrix} = \begin{bmatrix} r + D\frac{dr}{dx} \\ -1/f\left(r + D\frac{dr}{dx}\right) + \frac{dr}{dx} \end{bmatrix}$$

Figure 7.7 Matrices used in ray tracing examples.

Example:

Using the cavity geometry shown on the left in Figure 7.6, create a ray diagram matrix describing a single round trip pass through the cavity. Then plug in the following values: D=100 cm, R_1=50 cm, R_2=50 cm.

$$\begin{bmatrix} 1 & 0 \\ -2/R_1 & 1 \end{bmatrix} \begin{bmatrix} 1 & D \\ 0 & 1 \end{bmatrix} \begin{bmatrix} 1 & 0 \\ -2/R_2 & 1 \end{bmatrix} \begin{bmatrix} 1 & D \\ 0 & 1 \end{bmatrix} = \begin{bmatrix} 1 - \frac{2D}{R2} & 2D - \frac{2D^2}{R2} \\ \frac{4D}{R1R2} - \frac{2}{R1} - \frac{2}{R2} & 1 - \frac{2D}{R2} - \frac{4D}{R1} + \frac{4D^2}{R1R2} \end{bmatrix}$$

$$= \begin{bmatrix} 1 - \frac{200}{50} & 200 - \frac{2x10^4}{50} \\ \frac{400}{2500} - \frac{2}{50} - \frac{2}{50} & 1 - \frac{200}{50} - \frac{400}{50} + \frac{4x10^4}{2500} \end{bmatrix} = \begin{bmatrix} -3 & -200 \\ 0.08 & 5 \end{bmatrix}$$

These results represent a single round-trip pass through the cavity. Using this result, we could then step through several such round trips by multiplying this matrix by our ray vector (r_1, dr/dx), which we leave as an exercise.

Cavity Stability:

Using the above matrix manipulation techniques, it can be shown [6] that the cavity resonator is "stable" if:

$$0 \leq [1 - D/R_1]\,[1 - D/R_2] \leq 1$$

(where "D" is distance, and "R" is radius of curvature of the mirrors), otherwise it is "unstable", indicating that the photons generated within the gain media will eventually diverge out of the laser cavity.

Example: Cavity stability.

Using the dimensions in the previous example, is the cavity on the left in Figure 7.6 stable? (yes). If $R_1 = R_2 = 10$ D is it stable? (yes).

Exercise: Cavity stability.

Given the following set of cavity parameters, determine if the cavities are stable or unstable.

a) D = L = 10 cm, $R_1 = R_2 = 1000$ m.
b) D = L = 10 cm, $R_1 = R_2 = 4$ cm. (no)

Explain these results. Question: what caveat should we attach to the first case above? (Hint: consider the alignment).

Aperture:

The next significant element in the path of the wavefront is typically a shutter or aperture. This aperture limits both the amount of energy that exits the cavity, as well as defines the beam diameter. It can be as simple as a hole in the mirror, or may be as complex as a shutter which pulses open for a brief amount of time (e.g. ~micro seconds using "Kerr Cells"; see

Chapter 8) with the sole purpose of controlling the flow of photons from the cavity.

This exiting energy may or may not have a proper beam profile (e.g. beam diameter, beam divergence, etc.) for the intended application (e.g. coupling edge-emitting diode lasers into a fiber optic cable). Therefore, we typically attach some type of beam modifying optics at the output end of the cavity which either culminates or directs our beam as needed, such as a lens, or combination of lenses.

<u>Lens:</u>[35]

The most common and familiar type of lens is one made of shaped glass which is usually either curved in towards the center of the lens ("*concave*") or is curved out from the center ("*convex*")[36]. The lens effectively bends light as it enters due to the different index of refraction of the lens material compared to the media it is in (e.g. air), refracting according to Snell's law). By changing the radius of curvature of the lens surface, all the various wavefronts entering the lens at different points along its surface, can all be made to converge at some distant point

---

[35] At the risk of stating the obvious, we note that in addition to the calculations shown here, part of the cavity design process should include an analysis of the index of refraction of each lens *at the wavelength being used.* A glass lens that is transparent at visible wavelengths, may be opaque at other wavelengths (e.g infrared). Opaqueness not only reduces beam output power, it promotes heat absorption in the lens (i.e. damage).

[36] This is not the only type of lens structure possible. Another type which has become popular in the past 20 years is the Fresnel lens (pronounced "Fre'nel") which is made using a thin sheet of glass or plastic scribed with small groves or ridges. These groves/ridges promote constructive interference for a small range of wavelengths towards its focal point, thereby focusing light towards a favored direction, similar to a diffraction grating (see below). Since Fresnel lenses focus through diffraction rather than refraction (i.e. Snell's Law), they do not rely on a change in the index of refraction to work, which makes them ideal candidates for x-rays (the index of refraction of glass at x-ray wavelengths is ~1.0, which implies x-rays do not "bend" much in glass).

outside the lens. Together, the curvature of the lens surfaces and the index of refraction combine to yield the characteristic "focal point" of the lens ("f"). A *positive* focal point indicates the rays *converge*, while a *negative* focal point indicate they *diverge* as they exit the lens (e.g. a double concave lens):

$$1 / f = (n - 1) (1 / R_1 - 1 / R_2) \qquad \text{(thin lens approx.)}$$

where "f" = the focal point, "n" = the index of refraction of the material in the lens (e.g. crown glass = 1.53 at 590nm) and the "R's" are the radius of curvature of each face of the lens (by convention, the right side of the lens is positive, while the left side is negative). The relation of the focal length, object position and image position is:

$$1 / f = 1/D_{object} + 1/D_{image}$$

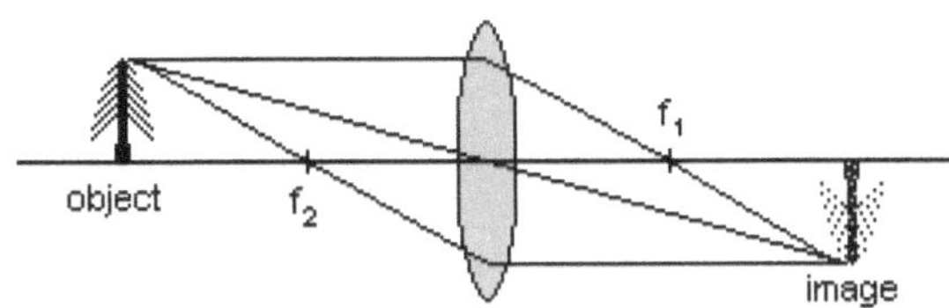

Figure 7.8 The lens effect.

Example: Lens focal point.

Using the above diagram as a guide, compute the focal length of a double convex lens with the following parameters: n = 1.53, $R_1 = R_2 = 50$ cm.

$$\begin{aligned} 1 / f &= (1.53 - 1) (1 / 0.5 - 1 / [-0.5]) \\ &= 2.12 \end{aligned}$$

Therefore the focal point "f" = 1 / 2.12 = 0.472 meters.

Example: Image position

Using the above lens, find the location of the image if the object is placed 1.5 meters away from the lens.

$$1 / f \quad = \quad 1/D_{object} + 1/D_{image}$$

$$1 / (0.472) \quad = \quad 1 / 1.5 + 1/D_{image}$$

therefore:

$$1/D_{image} = 1 / (0.472) - 1/1.5 = 1.45 \text{ meters}$$

which means the image will appear 0.69 meters from the lens (i.e. D = 1/1.45m).

Filters:

Assuming the cavity resonator is stable and well-aligned, we still need to deal with the fact that the laser cavity can support multiple resonant modes ($n = 2L/\lambda$), which implies our laser output may contain multiple spectral peaks. Since the gain media itself can generate both a spread in the wavelength as well as possibly have multiple resonant peaks, some method of removing all but the desired wavelength is needed, particularly in applications where poor spectral purity would prove detrimental (e.g. long haul fiber optic links – see discussion on Dispersion and Fiber Optics in Chapter 8).

If the energy we were generating were an electronic signal (i.e. transmitted by charged particles), we would simply put it through a frequency-selective "band-pass" filter made from components that affect its electrical behavior (e.g. capacitors and/or inductors). Since photons have no charge, we cannot simply use analogous components designed to affect electrical characteristics. How then do we separate the various optical frequencies out of our signal? By exploiting another, more fundamental aspect of the signal: its physical wavelength.

Most methods used to filter optical signals work by exploiting wavelength dependent constructive and destructive interference in the signal in order to select the desired wavelength while rejecting all others. One very effective method of doing this is to force the light through a "grating" made of a number of evenly-spaced slits. On the other side of these slits, the light recombines by Hyugen's principle to form a diffraction pattern. The resulting diffraction pattern is highly dependent on the wavelength of the light being used (as discussed under the topic of Young's Double slit experiment in Chapter 3), and is therefore very effective as a frequency selective filter.

A second method used to filter optical signals, is to construct a tuned filter composed of multiple layers of alternating optical densities (i.e. different indexes of refraction), with each segment spaced some multiple of $\lambda/2$ apart. Such a device is known as a "Fabret-Perot etalon". The etalon selects the desired wavelength by creating constructive interference as the signal reflects off the multiple faces inside its structure. All other wavelengths have the wrong dimensions to undergo constructive interference and eventually die out.

Diffraction Gratings:

The diffraction grating approach works on the same principle as did Young's Double slit experiment (see Chapter 3), only using more than just two apertures. As the number of slits ("N") in the mask (the "Diffraction Grating") increases, we find that the diffraction pattern cast onto the screen develops increasingly narrower and sharper diffraction peaks.

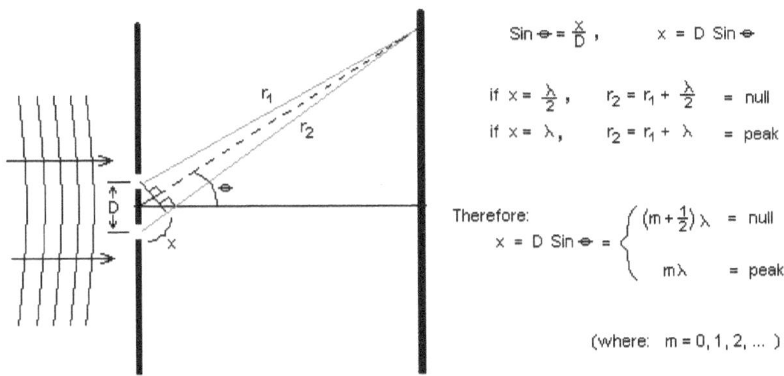

r_1
r_2
D
θ
x
$\text{Sin}\,\theta = \frac{x}{D}$, $x = D\,\text{Sin}\,\theta$
if $x = \frac{\lambda}{2}$, $r_2 = r_1 + \frac{\lambda}{2}$ = null
if $x = \lambda$, $r_2 = r_1 + \lambda$ = peak
Therefore:
$x = D\,\text{Sin}\,\theta = \begin{cases} (m+\frac{1}{2})\lambda & = \text{null} \\ m\lambda & = \text{peak} \end{cases}$
(where: $m = 0, 1, 2, \ldots$)

Figure 7.9 Young's Double slit experiment.

For example, consider a diffraction grating where N=15,000 slits on a 30mm face, the spacing between each slit is:

30×10^{-3} m / 15×10^{3} slits
= 2×10^{-6} m between each slit

If $\Delta\theta$ is defined as the angle between adjacent nulls around the central peak, we find:

$$D \; \mathrm{Sin}(\Delta\theta) = \lambda / N$$

By using the small angle approximation, this becomes:

$$D \; \Delta\theta = \lambda / N$$

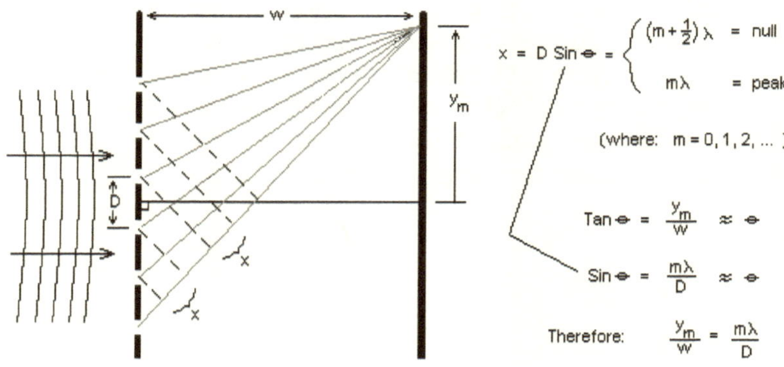

Figure 7.10 Diffraction grating, N=6.

Example: Diffraction grating

With λ = 632.8nm, N = 15,000 and D = 2×10^{-6} m, find the spacing between the nulls around the central maxima in degrees.

$$\Delta\theta = \lambda / (N\,D) = 632.8 \text{ nm} /(15{,}000 \;\; 2 \times 10^{-6} \text{ m})$$

$$= 21.1 \times 10^{-6} \text{ radians}$$

or:

$$= 1.21 \times 10^{-3} \text{ degrees}$$

In other words, by using a grating with a very large number of closely spaced slits, we will get a diffraction pattern where the bulk of the energy in the central peak covers no more than 1×10^{-3} degrees.

From the above figure, we see that the location of the "m^{th}" peak "y_m" from the central axis is a function of both the wavelength and the spacing between each slit:

$$y_m = m\,\lambda\,w / D$$

Example: location of 5th maxima

Assuming a width "w" in our system of 0.3 m, find the "y" location for the m=5 peak for two similar wavelengths, 633 nm and 635nm:

$$^{633}y_5 = 5\,(633 \text{ nm})\,(0.3 \text{ m}) \;/\; (2 \times 10^{-6} \text{ m})$$
$$= 0.47460 \text{ m}$$

$$^{635}y_5 = 5\,(635 \text{ nm})\,(0.3 \text{ m}) \;/\; (2 \times 10^{-6} \text{ m})$$
$$= 0.47625 \text{ m}$$

or 1.65 mm apart. This is more than enough separation to allow us to isolate one wavelength from the other if they were both present in the beam, even when those wavelengths are very close to each other (e.g. 2nm apart in the above example).

Should we need greater separation, we simply use a grating with a smaller "D" spacing.

For this reason, diffraction gratings are commonly used in applications where multiple spectral peaks need to be individually extracted (e.g. "multiplexed" fiber optic telecommunications links, where more than 128 different signals can be sent down one fiber by placing each signal at slightly different wavelengths. See Chapter 8).

Fabry-Perot etalon:

As mentioned, another very common method of filtering light by wavelength is to construct an "etalon" composed of multiple layers of materials with alternating refraction indexes, spaced some fixed-multiple of a half wavelength apart.

From previous discussions, we know that anytime energy propagating through a medium encounters a change in the propagation characteristics of that medium, some portion of that energy is reflected at the point where the characteristics change (see Appendix A).

If the index of the material it is entering is *greater than* the index of the material it is currently in (i.e. $n_{next} > n_{current}$), the reflected wavefront undergoes a 180 inversion in polarity at the interface (this is analogous to a pulse traveling on a thin rope connected to a heavy object, where the reflected pulse inverts at the interface to produce a null at the junction). If $n_{current} = n_{next}$, then there is no reflection. If $n_{next} < n_{current}$, then a reflection occurs, but it does not experience a 180 inversion[37].

If the spacing "D" between each layer is exactly one half wavelength, the reflections will all exit the right face of the etalon in-phase:

---

37 This 180° inversion in the Electric field polarity is due to the fact that the fields must be continuous through the interface. The nature of the interface itself is governed by the refraction indices, therefore: $E_1^{refl} = (n_2 - n_1) / (n_2 + n_1) E_1^{inc}$ which is negative if: $n_2 < n_1$.

$$\text{Reflection} = 2(\lambda/2)_{path} + [\,2(\lambda/2)_{path} + 2(\lambda/2)_{invrsn}\,] + 2(\lambda/2)_{path} + [2(\lambda/2)_{path} + 2(\lambda/2)_{invrsn}\,] + 2(\lambda/2)_{path} + \ \ldots$$

(where: $n_a < n_b$)

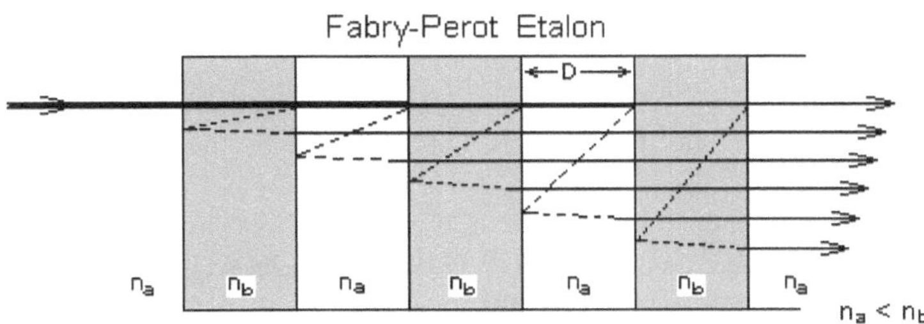

Figure 7.11 Fabry-Perot etalon.

As a result, all waves that meet the requirement D = nλ/2, *constructively* interfere through the etalon and in the process superimpose to reinforce each other as they travel. Wavelengths that do *not* meet this requirement *destructively* interfere as they reflect off the various faces inside the etalon. As a result, the etalon effectively forms a very selective *band-pass* filter which removes all signals but those at the desired wavelength (or integer multiples of λ/2).

(As an aside, we note that if you have ever seen the "rainbow" effect in a puddle of oily water, you have seen this effect in action. When the thickness of the thin sheen of oil on the surface of the water corresponds to a specific wavelength/"color" of light, it passes that wavelength. This effectively filters out all other wavelengths that do not correspond to the sheen thickness at that location. Likewise, other areas in the sheen with different thicknesses selectively pass different "colors".)

Assuming all energy injected into the etalon is conserved, the amount of energy transmitted and the amount reflected must equal the total amount of energy injected into the etalon (i.e. 100%):

$$T + R = 1$$

Noting that intensity is proportional to energy in the E&M field (where the energy in the field is the square of the magnitude of the Electric field, $I \propto |E|^2$), we can write this equation in terms of "intensity":

$$I_T / I_{in} + I_R / I_{in} = 1$$

In other words, the ratio of the energy (or "intensity") Transmitted vs. the energy injected, *plus* the ratio of the energy Reflected vs. the energy injected must equal 100%. As a result, *if we know how much is reflected, we automatically know how much is transmitted*, since the sum of $R + T = 1$.

We find that the "selectivity" of the etalon is directly related to the reflectivity of each interface: The larger "R" is, the greater the number of reflections within the etalon. The more reflections a "ray" undergoes within the etalon, the more it is either reinforced (if it is the correct wavelength), or nulled (if it is not the correct wavelength), and hence the narrower the "band-pass" response of the etalon (i.e. the more selective it is as a filter)[38]. This is characterized by its "finesse" "F":

$$F = \pi \, \mathrm{SQRT}(R) / (1 - R)$$

---

[38] Current fabrication techniques provide Reflectivities of 99% or better (i.e. $F > 300$).

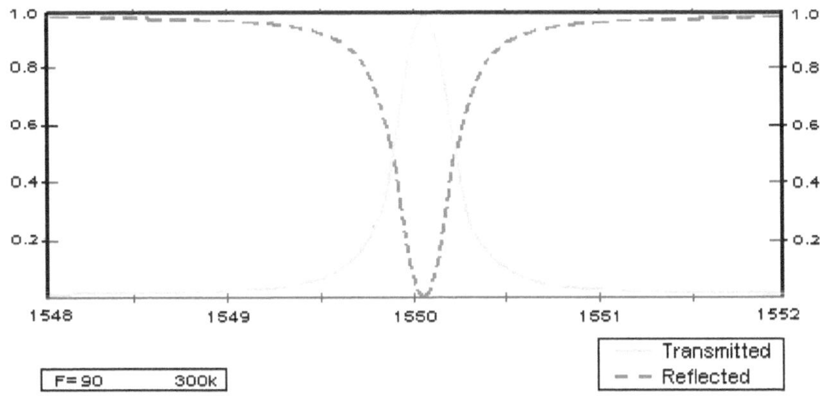

Figure 7.12 etalon I_T & I_R plot, Finesse=90.

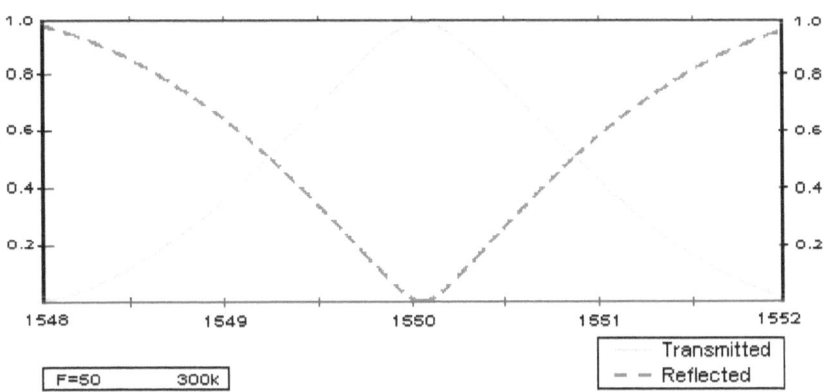

Figure 7.13 Etalon I_T & I_R plot, Finesse=50.

There are several different ways that these filters can be fabricated[39] within a section of fiber, including exposing the fiber

39 Though these older direct techniques are useful, they do tend to favor odd harmonics (analogous to our square wave example in Chapter 3). To address this bias, several newer fabrication methods have been developed which use a more *graduated* interface for a cleaner filter response (See article by Gubsky, Skorucak et al [19]).

section to high-intensity UV light through a chrome-silica mask, via etched mirrors, or through a micro-lithographic mask printed on the fiber itself.

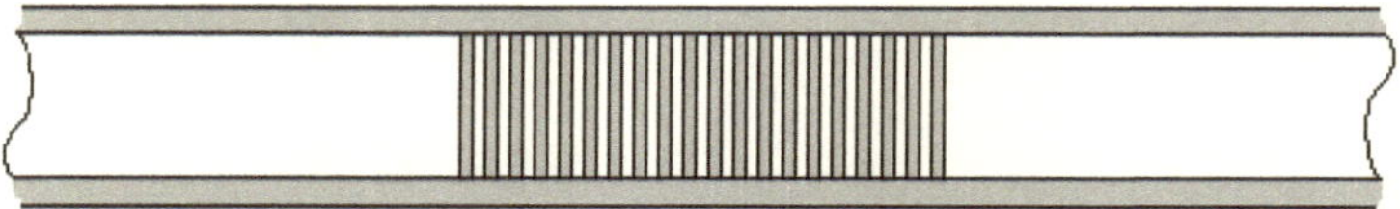

Figure 7.14 Fabry-Perot etalon imprinted in a Fiber segment.

Since UV light (e.g. 100nm) is at a higher frequency than visible (e.g. 600nm) or Infrared (e.g. 1300nm), UV photons have much more energy (recall that E = hf). As a result, UV light tends to erode molecular bonds as they impact the exposed fiber, thereby altering its index of refraction in the process[40].

Example: Fabry-Perot etalon spacing

Find the length "D" of the UV treated segments in the etalon required for a wavelength of 1300nm and a wavelength of 1400nm. The approximate index of refraction for these segments at 1300 is 1.35, and is 1.38 for 1400.

Since the wavelength is typically given as "free-space" length, and since the fiber effects the propagation of light as a function of wavelength (i.e. different wavelengths propagate at different speeds in different materials), we first need to find the

---

[40] Incidentally, this is why old glass jars or bottles left exposed to sunlight eventually turn purple. The same effect occurs if the glass is exposed to x-ray photons, only more rapidly, due to the fact that x-rays are at an even higher frequency than UV and therefore have even more energy. Unfortunately, x-rays are much more difficult to focus than UV or visible, since most materials have very low indexes of refraction at x-ray frequencies (~ 1.0), making it much more difficult to make a "lens" to focus or manipulate x-rays.

physical length of the wave inside the fiber for 1300 and 1400nm light using the index of refraction for each wavelength:

$$^{1300}\lambda = {}^{1300}\lambda_o \,/\, n_{1300} = 1300 / 1.35$$
$$= 963\text{nm}$$

$$^{1400}\lambda = {}^{1400}\lambda_o \,/\, n_{1400} = 1400 / 1.38$$
$$= 1015\text{nm}$$

Now that we have found the approximate lengths of each wave in the treated segments of fiber, we can find the approximate physical lengths needed for these segments in the etalon:

$$^{1300}D = {}^{1300}\lambda / 2 = 963\text{nm} / 2$$
$$= 482\text{nm}$$

$$^{1400}D = {}^{1400}\lambda / 2 = 1015\text{nm} / 2$$
$$= 508 \text{ nm}$$

Example: Finesse

Find the Finesse for two etalons with reflectivity's ("R") of 10% and 90%.

$$^{0.1}F = \pi \; SQRT(0.1) \; / \; (1 - 0.1) = 0.9$$

$$^{0.9}F = \pi \; SQRT(0.9) \; / \; (1 - 0.9) = 30.0$$

Brewster's Windows:

There is one final topic that should be mentioned before we move on, namely filtering the "Polarity" of light. The "polarity" or "polarization" of light indicates the orientation of the Electric field. Since the photons being generated by the atoms or molecules in the gain media tend to be generated with a random orientation, their polarization is also randomly oriented. In some applications, we may want our laser beam to be polarized in a single direction (for example, vertically, or horizontally polarized light). For those applications, a polarizing filter[41] is typically inserted in the beam path prior to exiting the cavity.

One of the more common polarization filtering techniques used in laser cavities is a glass window inserted in the cavity at what is known as "*Brewster's angle*". We know that when energy encounters any change in the media it is traveling through, some portion of that energy is reflected back. This is includes when the Electric field vector is oriented at some random angle to the interface between the two media, including at our slanted glass window. Consequently we find that when the electric field if it is not aligned parallel to the surface of this window, energy is reflected.

If the Electric field vector is nearly parallel to the plane of incidence, much less of its energy is reflected. When the glass surface is at *Brewster's angle* (a function of the index of refraction of the glass), the transition between the two media is very smooth, resulting in virtually no reflection at the interface for light polarized at this angle. As a result, anytime we insert a glass window at the exit of the cavity that is inclined at Brewster's angle, the emerging light is highly polarized along the plane of incidence.

$$\Theta_{\text{Brewster's}} = \text{ArcTan}(n)$$

(where "n" is the index of refraction of the glass).

---

41 Most sunglasses sold today for example, use such polarization filters to eliminate the glare being scattered off various surfaces. Since reflections scattered off horizontal surfaces tend to be horizontally polarized by the reflection, using sunglasses with filters aligned to pass only vertically polarized light tends to dramatically attenuate this non-vertically scattered "glare".

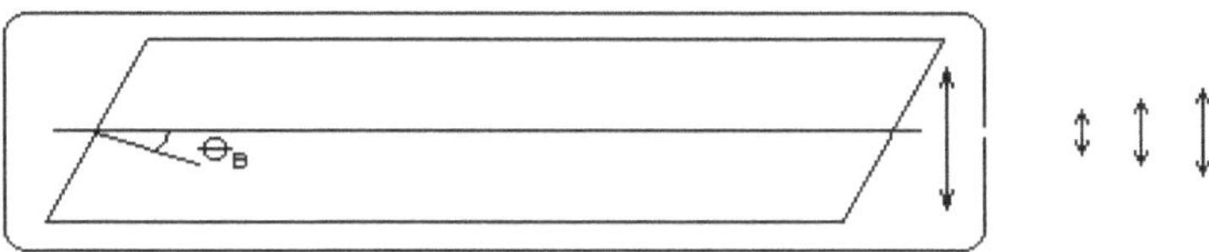

Figure 7.15 Cavity with Brewster's window.

Chapter 7 Review questions:

1. Why do we use mirrors around the gain media?
2. What happens when we enclose the gain media within a cavity?
3. Why is an integer multiples of the half wavelength a "magic" value in a cavity (or gratings, etalons, etc.)?
4. What advantages are there to using matrices to describe the components in the cavity?
5. What is meant by describing a lens or mirror as a "black box" transformer?
6. How does a grating filter light by wavelength?
7. How does a Fabry-Perot etalon filter light by wavelength?
8. What is the Finesse of an etalon filter?
9. How does a higher reflectivity at the etalon interface increase the selectivity of the etalon as a "band-pass" filter?
10. How does selectively exposing a fiber segment to UV light construct an etalon inside the fiber?
11. Explain how a thin sheen of oil on the surface of a puddle creates the "rainbow" effect (a selective filtering of "colors")?
12. Considering our Superposition example for a simple square wave in Chapter 3, why might we want to "soften" the interface between each segment of the etalon (Hint: Fourier theory question).
13. What is Polarization?
14. How does a Brewster's window selectively filter polarized light?
15. Returning to the example related to the cavity shown on the right in Figure 7.6 (given R1 = infinite), if R2 = D/2, is the

cavity stable? If R2 = D, is it stable? If R2 = 2D, is it stable? Explain the different results.

16. Given a double convex lens and Figure 7.8, calculate where the focal point is if $n = 1.5$, $R_1 = R_2 = 35$ cm. (0.35 m)

17. Given a fiber with $n = 1.4$, calculate the spacing needed to create a Fabry-Perot etalon for 850 nM. (303.6 nm)

18. If the reflectivity for the etalon created in the previous question is 95%, calculate its Finesse.

Chapter 8 Lasers and Fiber Optics

Information and Occupied Bandwidth:

When transmitting signals from point "A" to point "B", there is always some limit on the amount of information that can be carried for a given amount of time. This is true whether that information is sent via carrier pigeon, pony express, telegraph, phone line, pneumatic suction tube, RF channel, or fiber optic cable, etc. The transport mechanism, much like a water pipe, has a maximum "flow rate", and as we *approach* that maximum, the system begins to "saturate".

In the case of a copper phone line for example, the original purpose and goal of the system by design was to carry simple voice traffic. For maximum "bang for the buck" (the maximum number of users accommodated for a given amount of network system hardware investment), it was decided that all circuit components in the system would have a maximum upper frequency limit of about ~3.3 kHz. An individual may attempt to send frequencies higher than this, but the only guarantee "Ma Bell" will make for the normal circuit path is that it will pass up to a 3.3 kHz signal. Anything higher than that carries absolutely no guarantees whatsoever (and is typically "rolled off" by the limited band pass).

Though it is true that voice conversations can contain higher frequency components than this, virtually all of the information content in the vocal signal is contained well below 3khz during normal conversation, and therefore eliminating this upper spectral content represents a minimal loss of information content. To provide higher bandwidth than this for simple voice conversation, typically would be a waste of resources which could better be utilized elsewhere. In effect, this tradeoff enabled more users to place more calls in a given amount of time for an investment of "x" Dollars (or Pounds, or Deutschmarks, etc.) in the network, with only negligible impact on information flow. In essence, regardless of the technology used there are always competing requirements and *tradeoffs* that must be considered, defining something a fundamental part of the engineering that goes into building any such a network.

Over time people inevitably found ways to strain the throughput of the basic telephone system, initially through a simple increase in the shear number of voice calls placed during peak hours, and then as technologies developed, through higher throughput demands per connection (e.g. digital data file exchanges, internet browsing, video conferencing, etc.).

Anytime we transmit a signal through a given channel, we are effectively putting a fixed amount of energy into the assigned bandwidth of that channel (e.g. a 3 kHz bandwidth in the case of the typical phone system, or over an old cell phone channel, etc.). If instead of a simple voice transaction, we wanted to send high frequency content (e.g. a clean reproduction of a symphonic recording, or streaming analog video images, etc.), a 3 kHz signal bandwidth would be too narrow to pass the higher frequency components of that signal. Such broadband signals are often referred to as being "*spectrally rich*", or as having a "*high spectral content*", since the higher amount of "information" or content tends to contain much higher frequencies (i.e. spectral components) than a simple voice conversation.

In order to transmit higher "Spectral Density" signals, we needed to develop a network capable of providing the higher band-pass requirements than is available through the standard analog telephone voice transmission.

If for example we attempted to send a *real-time* analog video signal through our a low bandpass phone line, we would find that the amount of information contained in each picture could easily be far greater than the "band-pass" capabilities of a simple phone line. Television is a very spectrally rich signal, since it not only carries the information contained in the audio portion of the program, but also contains a vast amount of information contained in a rapid series of still pictures. Even one of these still pictures can be loaded with a variety of fine detail which corresponds to a tremendous amount of information.

For example, analog NTSC broadcast television signals sent 30 such pictures per second, each dissected into single lines traced horizontally across the screen at a rate of 15,734 lines per second (64μs per line). Each of these lines were themselves composed of thousands of dots (pixels), corresponding loosely to an equivalent of several Megabytes worth of information per frame. To accommodate this, the standard NTSC analog Broadcast TV channel was allocated 6MHz of bandwidth to send the combined Video and Audio spectral content – *information, or its energy, spread out across a given bandwidth*. If we attempted to force such a spectrally rich video signal through a 3 kHz or 30 kHz band-pass in "real-time", the resultant output signal would emerge from the system with such extremely severe *dispersion* and high frequency "roll-off", that virtually none of the fine detail or syncing information present in the original image would remain, effectively rendering the images as little more than blurred, torn, rolling *smears*.

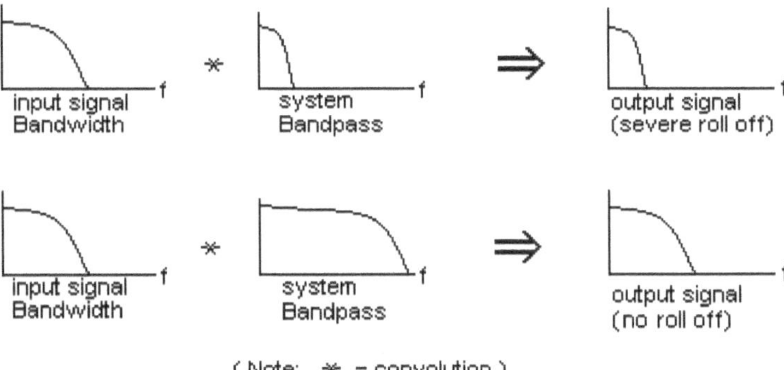

Figure 8.1 Effects of system band-pass on a signal.

A very good demonstration of this concept can be found in our discussion of the principle of superposition in Chapter 3, where we demonstrated what happens when we restrict or eliminate the higher frequency components in a square wave. In that example (see Figure 3.5), only the lower *three* frequency components were included, which resulted in a very rounded "square wave" with very long rise and fall times. In the second set of pictures, we included some of the higher frequency components, and the resulting waveform produced a much sharper square wave, with much shorter rise and fall times. We found that even though our square wave was running at 100 cycles per second, it contained higher frequency components that were well above 2,000 cycles per second, and which were essential to the "crispness" of our waveform. From this analysis, we see that anytime we restrict or eliminate the higher frequency components in a signal (for example, by forcing that signal through a system with too narrow a band-pass), the signal that emerges from our system will be severely distorted/degraded.

For example, losing the high frequency components that make up the rapid rise and fall edges in a square bit stream, effectively smears the bits into each other, producing Inter-Symbol Interference ("ISI"). Consequently the faster we try to force data through a low bandwidth system (i.e. the more we increase the Bit-Rate of the data processed through our telecom system), the wider our system band-pass needs to be to accommodate the higher frequency components. Otherwise our output signal suffers severe roll-off of the higher frequency components. As that happens, the bits to will begin to overlap, creating ISI in the process.

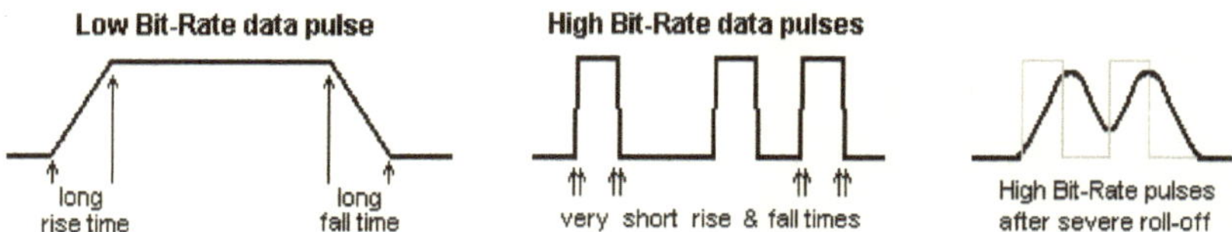

Figure 8.2 Slow Rate bit pulse with long rise times, compared to fast Rate bit pulse. High frequency pulses subjected to a low band-pass eliminates the higher frequency components of the pulses causing bits to overlap, producing Inter-Symbol Interference (ISI).

As we attempt to press for higher data bit rates and thus greater throughputs (i.e. more data across a fixed signal path in a given amount of time), we find our telecommunications system needs to have ever-greater bandwidths. Though there are some tricks that we can use to allow us to squeeze more data into a given channel width (e.g. using Amplitude Phase Shift keying "APSK"), these techniques increase the complexity of the system while increasing the sensitivity of the signal to noise. To compensate for that increased sensitivity to noise, we need to increase the signal strength (i.e. inject more signal energy into the BW) to overcome the noise. As more users increase their signal strength, the more they inadvertently inflict more noise on each other (particularly in shared or adjacent systems).

In other words, though we can find ways to increase the amount of information we can pack into our channel, we know from the Law of Entropy that in order to achieve this we must pay a "higher price" – both in power, and in the increase in complexity of the mechanism we employ in our system to overcome the effects of Entropy. In a channelized RF ("Radio Frequency") system for example, a simple doubling of system throughput can often cost us several orders of magnitude in complexity (and hence Dollars). However if we are aware of the pitfalls and are willing to pay the price to overcome them, Entropy can be checked (on a local scale), at least *temporarily*. But eventually we reach a point where there is only so much you can do with the existing technology, requiring us to implement an entirely different approach in order to expand our limits.

The Fiber Optic link:

To meet the exploding demand, the old "copper-based" phone system had to undergo a dramatic change. This change came in the form of the introduction of glass fiber optic links as the new pathways needed to meet all that growing demand for bandwidth. In addition to developing the ability to manufacture *extremely* pure glass fibers 100 kilometers (65 miles) long in large quantities, we also needed to improve our ability to fabricate large numbers of infrared lasers needed to drive all those fiber optic links. Combining what we learned in our quest towards reaching the moon (including semiconductor fabrication), with on-going research in telecommunications, the technology of fiber optics and lasers developed enough by the 1970's to produce the first practical fiber optic link. These optic links effectively increased the bandwidth of a single "phone line" from ~3 Kb/s, to over 1 Gb/s (roughly a *million*-fold increase)!

In the decades that followed, the art of fabricating pure optical fibers and high power semiconductor lasers at various IR wavelengths at economical costs, as well as improvements in all the components related to the optic link, has helped to further expand the total system throughput *exponentially*. This in turn has triggered innovations in communications that have fueled even more demand far beyond what was envisioned even as recently as the 1990's. To keep up with that exponential increase in demand, recent developments in optical Wavelength Division Multiplexing ("WDM") have now enabled us to take a single fiber and send more than *fifty* times the volume of traffic that we were able to send on that same fiber just *twenty* years ago (see discussion on WDM which follows).

Typically, a simple fiber optic link is composed of a nearly monochromatic laser (e.g. 1310 nm semiconductor diode) fed a string of digital electrical "bits". The laser converts these digital bits into pulses of light which are then launched down a pure glass fiber slightly larger than a single strand of horse hair. Since the wavelength of the light is in the hundreds of nanometers (or conversely frequency is in the hundreds of Terahertz), extremely high frequency data signals ($>10^9$ bits/sec) can be used to modulate the laser carrier, and still not "shift" the laser wavelength more than a fraction of a percent.

The primary reason for the significant improvement in bandwidth between the fiber optic link and its older copper wire predecessor, is found in the linearity of Maxwell's Equations. Since the vehicle used to carry the information over the optic link is the photon itself (as opposed to the electron in copper wires), the fact that the photon has no charge, and the fact that the equations governing the propagation of photons are linear equations, implies these photons do not "mix" with other E&M photons/fields in the environment (i.e. noise) as they travel. Consequently the *ocean* of electromagnetic noise that we swim through our entire lives does not contaminate the information carried by the photons as they propagate down the optical fiber link.

In other words, when the information was carried by a charged particle (e.g. an electron), any stray E&M field (i.e. random noise) passing the electron would exert a Lorentz force on it, and thus couple its random noise into the signal the electron was carrying. Photons, having no charge, are impervious to such external E&M field influences, and are thus unaffected by them despite their ubiquitous nature.[42]

As in all things however, Entropy tells us there is "no free lunch". In designing and implementing such a system, special care must be taken to minimize the persistent effects of all this random noise, as well as the other degrading effects that Entropy guarantees lurks around every corner. In addition to random noise corruption in the electronic processing that feeds each fiber link, we must pay particular heed to the purity of the glass fiber used, as well as the "*spectral purity*" of the light injected into the fiber medium. Since the fiber's characteristic behavior depends heavily on wavelength, if we use a light source that contains a wide range of wavelengths or modes (for example), our ability to send and receive information over each fiber link will be significantly degraded, as we shall soon see.

Optical properties of Fiber waveguides:

Fiber optic cables are fabricated by extruding extremely pure glass stock "Preform" (Silicon Dioxide, doped with various amounts of Germanium to increase the index of refraction) into very fine thread[43] "*cores*". These cores are then coated by a "*cladding*" layer with a slightly lower "optical density" (i.e. lower index of refraction), forming an optical waveguide. When we launch a pulse of light down the fiber, the glass waveguide inevitably "transforms" our pulse of energy in three significant ways, namely:

1) Attenuation,

---

42 One could argue that copper wires also limit bandwidth through stray capacitance and inductance which couples signal energy to ground as well as induces counter EMF (see Chapter 6). However these are all electromagnetic field effects. Eliminate the field effects on the information carrying vehicle (the photon), and they cease to be an issue.

43 Single Mode fiber have diameter around 10 microns, while multi-mode fibers are considerably larger (roughly 50-100 microns in diameter).

2) Reflections, and
3) Dispersion.

Attenuation is simply the absorption or loss of energy as it travels down the glass fiber (e.g. by impurities). This loss of energy can be modeled fairly accurately using a simple exponential function (see Appendix C for a more rigorous treatment):

$$I_{(x)} = I_o \exp(-\alpha x)$$

where "I_o" is the initial launch intensity, "α" = the characteristic absorption of the fiber per km, and "x" = the distance down the fiber. The absorption or loss may also be expressed in dB (e.g. 0.35 dB/km at 1310 nm[44]). Obviously, in order to send the light pulse the greatest distance down the fiber, we need minimal loss of energy, and therefore as pure a glass fiber as possible.

In addition to absorption by impurities, there are several other mechanisms that contribute to loss in a fiber waveguide. These include: energy which escapes into the outer cladding layer (e.g. due to surface pitting or irregularities on the surface of the core, or sharp bends in the fiber which cause the angle of incidence to drop below the critical angle, causing a reduction in the amount of light experiencing Total Internal Reflection – see discussion below), as well as scattering off impurities in the glass. As one might expect from our discussions thus far, most of these losses tend to have a wavelength dependence to them.

For example, when a fiber is bent around a sharp edge, we find that shorter wavelength light tends to "negotiate the bends" much better than longer wavelength light, and hence the longer wavelengths (e.g. 1550nm) tend to be attenuated much more by bends in the cable than shorter wavelengths (e.g. 850nm). Energy scattering off impurities or defects (sometimes referred to as "Rayleigh Scattering") is also a significant wavelength dependent loss mechanism and is inversely

---

[44] By comparison, we point out that absorption rate for normal consumer grade "window" glass is at least two orders of magnitude higher.

proportional to the fourth power of the wavelength: $1/\lambda^4$. As a result, shorter wavelength light suffers more scattering loss than longer wavelengths (e.g. 850 nm suffers ~16 times more scattering loss than does 1610nm).

Finally, we know that water molecules (actually the OH portions of H_2O) tend to have a strong resonance at 950, 1383nm and above 1600, creating absorption peaks around these wavelengths, particularly if the glass has a high OH^- content (e.g. as an impurity that was introduced during the fabrication of the fiber). Improvements in the manufacturing of fiber strands have allowed manufacturers to eliminate the 1383nm absorption peak, thereby increasing the range of usable spectrum space from ~850nm to ~1600nm, easily doubling the amount of usable wavelengths, and thereby enabling a significant increase in the number of available WDM channel slots (see below).

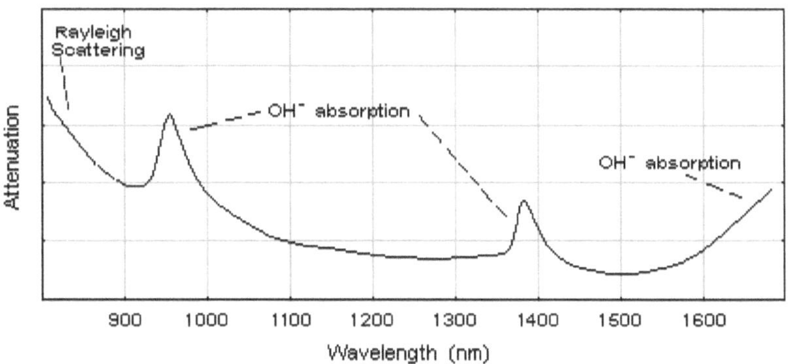

Figure 8.3 IR water absorption peaks affecting fiber optic links.

Reflections, on the other hand, occur in two general forms: The first is the result of the fact that some portion of energy is always reflected anytime a wave encounters a change in the propagation characteristics of the medium it is in (in this case, the glass fiber). This could be due to a physical defect in the fiber, a poor connection/splice, or simply due to the microscopic variations in the amorphous glass structure, which produces what is known as "*backscatter*" as the energy encounters these slight variations in the uniformity of the media.

The second type of reflections are those which occur each time the energy encounters the inside wall of the fiber. This is often depicted using the simplified approximation of ray optics, where the wavefronts are represented by a "ray" pointing in the direction of travel. As these "rays" meet the edge of the fiber wall at a glancing angle, they undergo "Total Internal Reflection", as described in Chapter 5. This is governed by Snell's Law of Refraction:

$$n_1 \sin(\theta_1) = n_2 \sin(\theta_2) \qquad \text{Snell's Law}$$

By rearranging terms we find:

$$\theta_2 = \text{ArcSin}\,[\, n_1 / n_2 \,]$$

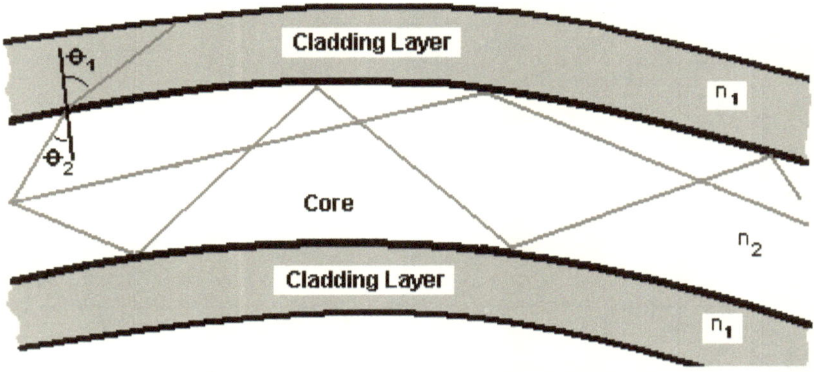

Figure 8.4 Total Internal Reflections off optical fiber walls. When the angle of incidence of a ray is less than the critical angle, it does not experience Total Internal Reflection, and thus exits the core and enters into the cladding.

Since the fiber is manufactured specifically such that the inner core of the fiber has a higher index of refraction than the cladding layer surrounding the fiber, Total Internal Refraction will not occur for any "ray" of light that strikes the outer edges of the fiber at any angle less than the critical angle (see Chapter 5):

$$\theta_{critical} = \text{ArcSin} [n_1 / n_2] .$$

<u>Chromatic Dispersion</u>:

From our previous discussion, it should be clear that though the index of refraction of the basic optical fiber is relatively constant throughout the fiber, it is not constant in terms of wavelength. This wavelength dependence on the index of refraction is inherently due to the intrinsic response of the atoms in the fiber (see Chapters 3 and 4). This results in a different attenuation, absorption and/or delay being applied to different wavelengths moving through the fiber, leading collectively to "Chromatic Dispersion".

A common example of Chromatic Dispersion occurs when light refracts through a prism or raindrop. Since each of the different wavelengths (or "colors") have slightly different refraction indexes, each wavelength is bent at a slightly different angle, separating each wavelength out across a range of angles. As a result of the variation in the index of refraction for different wavelengths, if we use a laser that is *not* monochromatic (e.g. a laser with a multi-peak profile), each of these various wavelength components will travel down the fiber at slightly different speeds, emerging from the end of the fiber at slightly different times.

When these components emerge at the other end of the fiber, some will be slightly leading, while others will be slightly lagging the others. When this collection of staggered pulses is detected by the receiver (typically a photodiode which converts the light back into an electrical pulse), it will simply sum all this energy together indiscriminately. This resulting sum, therefore will *not* be identical to the original pulse injected at the transmitter, but instead will be something of a "smeared" version of it, resulting in ISI.

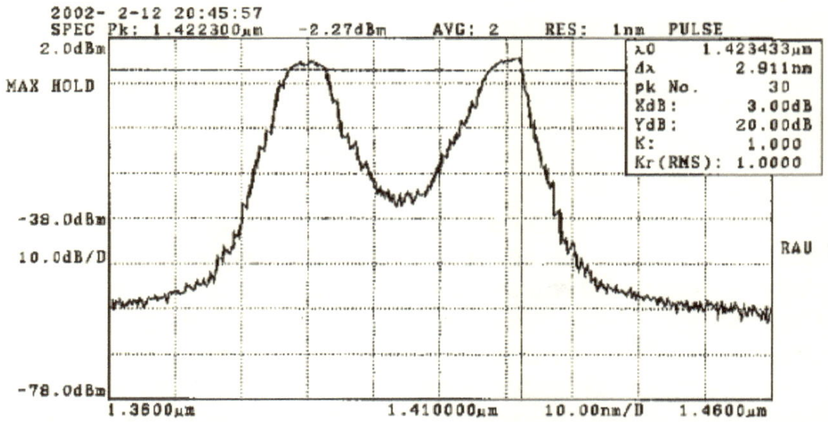

Figure 8.5 The spectral profile of an average semiconductor laser, exhibiting multiple wavelength peaks.

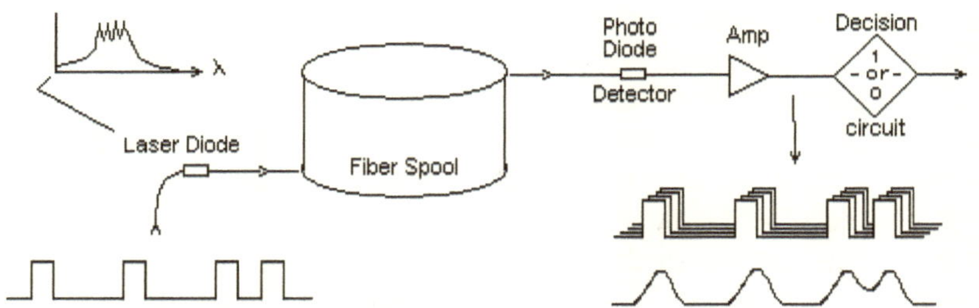

Figure 8.6 Dispersion effects on digital pulses. Multiple versions of the signal's bit pulses arrive at the detector slightly out of sync and are summed by the detector into one lumped composite. If the bit pulses are widely separated (i.e. low rate data), their "smeared" version at the output will not overlap. If they are close together (i.e. high rate data), the smeared pulses will overlap, creating ISI.

If our data rate is slow (i.e. plenty of time between each pulse), each received pulse will remain distinct enough to be easily isolated and identified by the detecting circuitry at the receiving end. If however, we want to send high rate data (i.e. we want to increase the system throughput by sending the bits at a faster speed), we eventually run into a problem where these "smeared" bits begin to overlap. At that point, the receiver electronics will begin to misread some of the adjacent bits, designating what should have been a "0" as a "1" (and vice versa) – leading to "*Inter Symbol Interference*" ("ISI"). Though there are various techniques that allow us to do some amount of error correction, when these bit errors become too severe, all error correction techniques eventually become overwhelmed and begin to break down. As a result, this effectively forms the upper bound on the throughput of our system.

It is possible to offset the effects of dispersion by using what are known as "Dispersion Compensators" and "Dispersion Shifted Fibers". These devices are specifically fabricated such that they have exactly the opposite frequency response of the fiber they are attached to, thereby performing an "inverse" dispersion transform on the signal. This inverse effect equalizes the delay by similarly affecting the higher frequency components to compensate, and thereby allowing the lower frequency components to "catch up". These compensators are typically designed with specific wavelengths in mind (e.g. a positive shift above 1550nm, and a negative shift below 1550nm, etc.).

Effects of wide Launch angles:

Another problem we encounter in sending information over a fiber link occurs when the light source used has a wide divergence angle (see our previous discussion on semiconductor edge-fired laser diodes in Chapter 5). If the energy launched into the fiber has a wide range of launch angles, the light "rays" entering the fiber at *steep* angles will undergo far more reflections than those that enter the fiber straight in. As a result, these steep angle "rays" will travel a greater distance down the fiber (compared to the shallow angle rays), and hence will exit the fiber later than those rays launched straight into the fiber core (see Figure 8.4). Consequently any digital bits fed into this laser, will produce a staggered train of light pulses at the opposite end of the fiber. These will be indiscriminately summed by the detector, producing a degraded "hump" rather than a nice square pulse as was originally fed to the transmitting laser. These smeared pulses eventually begin to overlap with adjacent pulses in the more severe cases (e.g. longer fiber lengths, or higher bit rates), creating ISI.

In an attempt to minimize this effect, many fiber manufactures have developed a number of techniques, including fabricating the fiber with a "*graduated index of refraction*", where the index of refraction decreases with radius (either gradually or in several relatively abrupt steps). The lower index of refraction at the edges of the fiber means rays which spend more of their time traveling through the edge of the fiber will travel faster than those which spend most of their time traveling down the center of the fiber. This effectively compensates for the time difference between the rays launched at a steep angle compared to those launched down the center, reducing the time differences between the two paths, and thereby minimizing ISI.

Multi "mode" vs. Single "mode" fiber:

A third condition that also contributes to dispersion and thus ISI in long cables, occurs when several propagating "modes" exit the laser cavity at the same time (see Chapters 5 and 6) and are allowed to couple into the fiber. Since the fiber itself represents a waveguide, its size and shape has a dramatic impact on how these modes propagate down the link.

From our discussions in Chapter 6, we know that there is a maximum wavelength that can be supported by the structure (the fundamental mode, corresponding to the "cutoff" frequency). All wavelengths that are longer than this (i.e. frequencies that are below the cutoff frequency), will not be supported by the waveguide and consequently will die out rapidly. How this energy distributes across the diameter of the fiber as different modes, determines how it propagates down the fiber. For example, the fundamental mode TEM_{00} (represented by Bessel Function J_o in our discussion in Chapter 6) propagates primarily down the *center* of the fiber, while the TEM_{11} mode tends to be symmetrically offset from the center of the fiber (etc.), and thus propagates differently than the TEM_{00} mode.

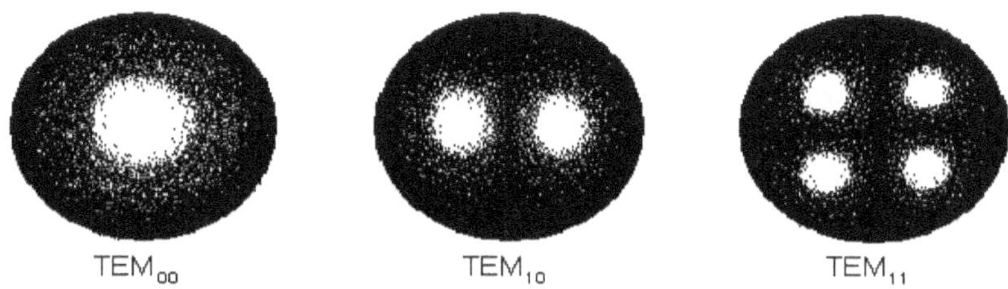

Figure 8.7 Several possible TEM modes across the face of a fiber

Since these different resonant modes propagate through the fiber differently (i.e. experience a different "black box" transform processing by the fiber), they emerge out the other end at slightly different times. If the fiber is short, the time difference is minimal, and ISI is not a problem. However, if the fiber is long, the time delays accumulate and ISI becomes a much more significant limitation on the data rate of the system. As a result, long haul optic links use very narrow fibers such that only the simplest mode can be launched down the fiber (i.e. restricting the diameter of the fiber effectively filters out all but the zeroth mode, since the waveguide is only wide enough to support this one central mode). Multi-mode fibers on the other hand, tend to be much wider and thus allow the additional modes (and hence more energy) to propagate down the fiber. As long as the fiber lengths are kept to less than say a few kilometers, the multi-mode dispersion / ISI is minimal. Consequently, multi-mode fibers are typically only used on shorter runs (e.g. inside a single building, or to a facility next door, etc.). Single Mode" fiber diameters are around 10 microns, while multi-mode fiber diameters range from 50 to 100 microns.

Fiber Amplifiers:

When shooting a laser pulse down a long length of fiber, absorption and scattering losses eventually diminish the field intensity down to where it is no longer usable. If the signal being sent is expected to travel any great distance (i.e. > ~100km, or 65 miles), a "repeater" station is required to boost the signal before it has attenuated below a usable level. In the past, repeater stations were composed of a simple detector, electronic amplifier and laser transmitter. The detector received the original signal and converted it back into an electrical signal to be filtered, cleaned and re-shaped by the electronics. After reshaping the pulse, another laser transmitter then re-launched it on down the fiber towards the next repeater.

Though such repeater systems were functional, they tended to introduce random E&M noise (via the electronics), and added complexity (i.e. increased number of failure modes) to the link that system designers wanted to eliminate where possible. The preferred approach to all engineered systems is to minimize both the number of components and overall complexity of the system. In a fiber link, keeping the signal entirely in the photon realm would eliminate the light-to-electron processing step(s), while helping to minimize the component count as well as system complexity. Operating entirely at the optical level, also eliminates noise being induced into the signal by stray E&M fields (see previous discussion at the start of this chapter).

One method of achieving a system that operates purely in the optical realm, is to amplify our weak signals while still photons, by passing them through a pumped gain media at the amplifier station. Since the first two terms in the acronym LASER is *Light Amplification*, this approach to boosting the signal effectively extends the original laser [amplifying] cavity to the far end of the link.

Several solutions designed to achieve this have been devised using a specially doped fiber segment as a gain media. This doped fiber segment is driven by an external energy pump (e.g. flash lamp) which places the dopant atoms in this segment into a metastable state. These excited atoms then undergo stimulated emissions anytime photons exiting the passive fiber link enter the doped fiber segment, stimulating a coherent avalanche emission out of the treated segment. These photons are then re-launched down the passive section of the link towards the next repeater where the process is repeated several more times until the original signal reaches its final destination.

One of the more effective fiber amplifiers developed so far are those which are doped using the rare earth element Erbium ("Er", Atomic number 68) and are therefore known as "Erbium-Doped Fiber Amplifiers" (EDFAs), operating in the 1550 nm range. Alternate designs use the rare earth element Praseodymium ("Pr", Atomic number 59) as a dopant and are known as "Praseodymium-doped Fluoride Fiber Amplifiers" (PDFFAs). These operate in the 1310 nm range.

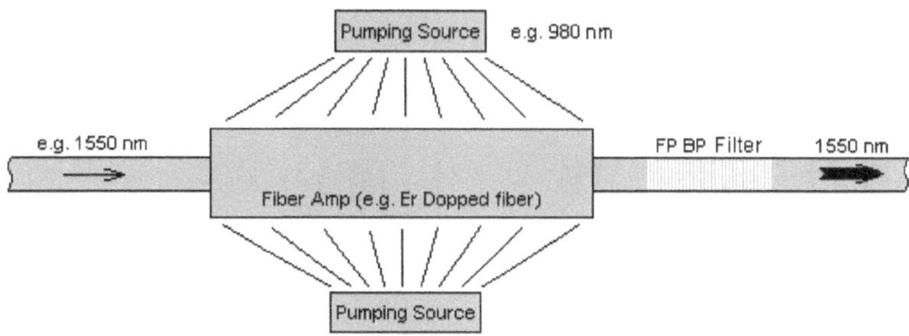

Figure 8.8 Er3+ doped Fiber Amplifier.

In the case of the EDFA, the doped segment is pumped by an external laser operating at ~800nm or ~980nm, which elevates the dopant (Erbium ions, Er3+) into their key metastable state. As the stimulated ions begin to de-excite, they emit photons at around 1545nm, which are then launched down the fiber. Prior to launching the signal however, it is passed through a 980nm filter to remove any traces of the pumping energy from the output beam, and thus remove this as a possible source of noise or dispersion induced ISI at the receiving end. The advantage of pumping the fiber at a wavelength well removed from the transmit wavelength (e.g. 1550nm for the EDFA), is that the pump energy does not introduce any unwanted energy/noise at the transmit wavelength.

Erbium infuses into silica relatively easily, making it a simple task to introduce it into the fiber during fabrication. Though the metastable state of Erbium is relatively long lived (~10 ms), it can be further increased by introducing GeO_2-Al_2O_3, which nearly doubles the metastable lifetime, thereby improving the Quantum efficiency of the lasing effect.

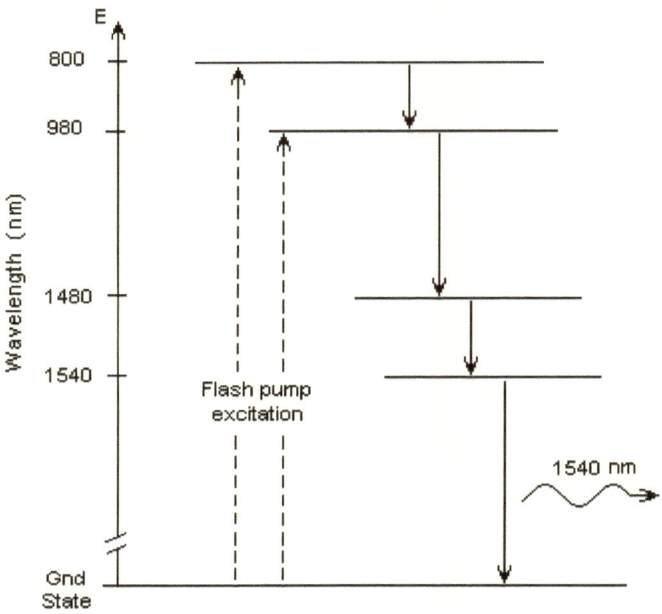

Figure 8.9 Energy States of Erbium.

<u>The Raman Effect</u>:

Though these rare earth amplifiers are very effective across their bands of operation (1550 nm for EDFAs, and 1310 for PDFFAs), their gain wavelengths are fixed by the energy levels of the ions themselves. To enable a wider choice of wavelengths, an alternate amplifier using the Raman Effect has been developed. The Raman "scattering" effect itself is based on a relatively simple concept, but like many of the topics in Quantum Physics, it cannot be explained using Classical models.

In Chapter 6 we introduced Maxwell's equations, and pointed out that they were linear equations, which implies electromagnetic energy does not mix or combine when it propagates through a "normal" (i.e. passive) medium, such as the vacuum of free space. We pointed out however, that electromagnetic energy does propagate differently in different media, which modified behavior we accounted for by including the "permittivity" and "permeability" of the media (something that typically varies with frequency). We stated that the frequency dependence of "ε" and "μ" is a consequence of the fact that different atoms or molecules within the media respond differently at different wavelengths.

In Chapter 4, we found that the resonant modes within atoms and molecules are what make it possible for the atoms and molecules to both selectively "tune" and extract out the desired wavelength from the incident pump energy we inject into the laser cavity. We mentioned that the storage of the energy key to lasing was via the metastable state, which allows us to establish the much needed population inversion in order to make lasing possible.

When the energy of the electromagnetic wave happens to correspond to a resonant frequency of an atom or molecule in the media (i.e. when the E&M photon's energy $E = hf$ equals the energy difference between two atomic or molecular energy states), that photon is absorbed. Now that the atom or molecule has absorbed this energy, what does it do with it? There are several possibilities.

The most probable response (i.e. the most common) is that the atom/molecule responds "passively" by immediately de-exciting (within $\sim 10^{-15}$ seconds) and in the process, re-radiates this energy by releasing a photon at the same frequency (but not necessarily in the same direction). This process in referred to as *Rayleigh Scattering* (or in classical parlance, as "elastic scattering", since no energy is lost). We mentioned in Chapter 4 that this scattering mechanism is responsible for the sky being blue, and why sunsets are mostly red/orange.

The next most likely de-excitation mode is known as *Raman*[45] *Scattering* and occurs when the atom/molecule de-excites into a level *other* than the level from which it started. In reality there are two distinct types of Raman emissions, known as *Stokes* and *Anti-Stokes* emissions.

If the atom de-excites to a valid level that is somewhere between the excited state and the starting level, it releases a photon that is at a *lower* frequency than the injected photon (in classical parlance, this would be considered an "in-elastic scattering" process, since the released photon has *less energy* than the original photon). This type of Raman scattering produces a *Stokes* spectral line and is the most probable type of Raman emission (since atoms or molecules found in nature tend to be mostly in the lowest energy state possible, as predicted by the Law of Entropy).

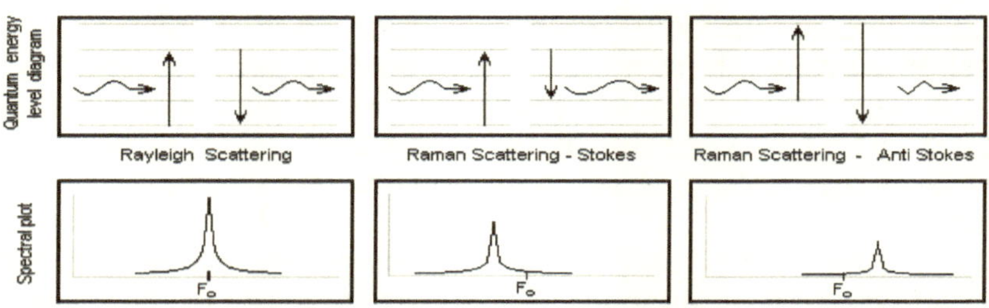

Figure 8.10 Rayleigh and Raman Scattering, with Stokes and Anti-stokes photons.

If on the other hand, the gain media atom was already in an excited state from a previous excitation event *prior* to the encounter with our signal, then our impinging signal's photon could trigger a de-excitation that can cause the atom to drop to a level lower than it started. If that occurs, the atom will release a photon that is higher in frequency than our initial pumping photon. The higher energy photon that is released in the process produces what is known as the *Anti-Stokes* spectral line. Clearly

---

45 The Raman Effect is named after Chandresenkhara Raman, who first observed it in 1928, five years after it was predicted by Adolf Smeckal.

this process is most definitely *not* a "passive" or linear process, and therefore there is no classical terminology to describe it.

Raman Amplifiers:

By exploiting the Raman Effect, scientists have developed a unique class of amplifiers, known as Raman amplifiers. In developing "photonic" amplifiers, one thing we do *not* want to do is pump the gain media using the *same* wavelength we are trying to amplify, since that would result in the pumping energy "drowning out" the weak signal energy we are trying to amplify. What we need is a pumping scheme that will allow us to excite the gain media at one wavelength, and then stimulate an emission at another. The Raman Effect offers us precisely such a mechanism.

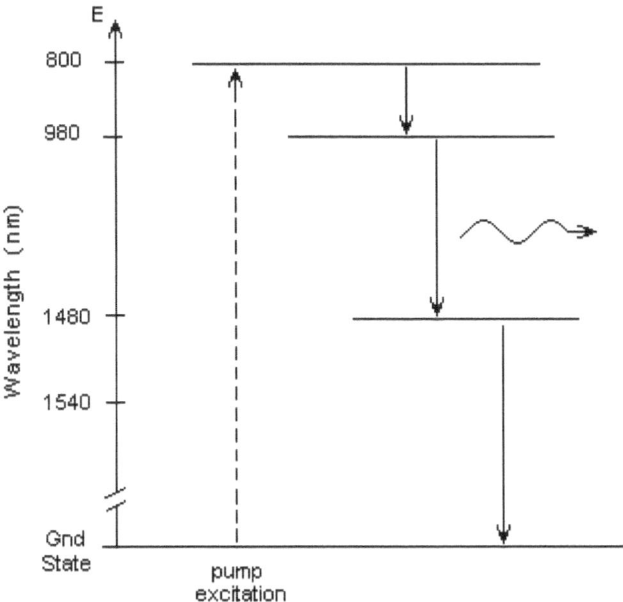

Figure 8.11 Raman Amp energy level diagram.

If we could setup the Raman Effect in a section of fiber, then it would be possible to use it to amplify incoming light at one wavelength by pumping the gain media at a different wavelength. This technique offers us a very attractive approach to amplifying weak signals without having to first "down-convert" everything to an electrical signal, amplifying that, and then "up-converting" this back to the optical realm once again using another laser transmitter (see previous discussion on EDFAs). By exploiting the Raman Effect, we can create an optical amplifier that will allow us to avoid the down/up conversion process, and all the related noise and contamination that comes with it.

Though conceptually similar to the rare earth amps, the gain wavelengths of Raman amplifiers are not dependent on the dopant ion resonances themselves. Instead the amplified wavelength processed by Raman amps depend on the pumping laser's wavelength. This enables Raman amplifiers to be used to amplify a much wider range of wavelengths than is possible using rare earth amplifiers.

One caveat in the use of such optical processing is that Doped Fiber and Raman amplifiers only allow us a way to amplify our signal strength. If we want to use the information in our signal to help route or "clean up" (etc.) our signal, current technology still requires us to down-convert everything to the electronic level first. If for example we want to use part of the bit stream "header" information to help route, or check parity, or otherwise perform logical ("AND"/"OR") processing on our signal, we must down-convert it into an electronic form to perform this level of processing (see discussion at the end of this chapter on Photonic Logical Processors).

Wavelength Division Multiplexing (WDM):

In our previous discussion about the legacy aspects of telephone systems, we mentioned that as time progressed, the sheer number of users attempting to place both local and long distance calls increased dramatically in the past three decades. In an attempt to accommodate all those users, the telephone company developed a radio link technique in which it relayed thousands of long distance calls across the country via microwave transmitters. These radio links first took each conversation and assigned it a single "slot" in an RF channel scheme. Each of these channels were then combined with hundreds of other conversations, each in their own designated "slot" of the RF microwave channel. This group of channels were then combined with still other groups (each in their own RF segment), amplified, and transmitted *en masse* out over the air towards their destination.

This process of "bundling" individual user connections together into one very large block of segmented "channels" by assigning each user a specific slot in the available RF spectrum is known as "*Frequency Division Multiplexing*". Other forms of Multiplexing exists, including "*Time Division Multiplexing*" (where each conversation is compressed into a short burst and sent during designated time "slots") and "*Code Division Multiplexing*" (where each user's conversation is convolved with an orthogonal digital code and then sent on the *same* RF channel along with all the other coded conversations – see our discussion on Cell Phones at "EpiphanyBySteveLee.com", misc tab for additional information). Since resources are always finite (be they RF channel space, copper phone lines, or fiber optic cables buried in the ground, etc.), it is essential that these resources be made as efficient as possible, and constructing a multiplexing scheme similar to those just described is one very effective solution[46].

The exponential increase in voice and then data traffic in recent decades has spurred a similar capacity demand from existing fiber links. To address this growing demand, fiber optic system designers have adopted an approach similar to the "Frequency Division Multiplexing" scheme described above. This approach, known as "*Wavelength Division Multiplexing*" (WDM), was first developed and implemented in the 1980s

46 Note that most systems in use today first digitize each conversation to enable the traffic on that connection to be processed more efficiently both before and after the multiplexing step. This processing includes a variety of bit manipulations such as parity checking or other enhancements to both increase the channel efficiency as well as lower the number of bit errors.

(using 850nm and 1300nm wavelength lasers). In WDM, each individual digitized bit stream is used to modulate the intensity of a separate laser at its own unique wavelength (e.g. 1305nm, 1310nm, etc.).

As a result, each individual bit stream is modulated up to the assigned wavelength much the same as if it were being assigned a radio channel slot in our microwave link scheme. Other user bit streams are similarly placed into unique wavelength "slots" adjacent to the first. This summed composite array of multiple modulated wavelengths is then launched down a single fiber towards its destination, allowing one optical fiber to carry *many* different user's traffic simultaneously.

At the receiving end, a filter (such as the gratings described in Chapter 7), is used to separate out each of the individual wavelength "channels", routing each individual wavelength to a separate detector/amplifier circuit for routing/processing. Using this WDM scheme, as many as 128 or more separate bit streams can be "piggy-backed" onto a *single* fiber, thereby greatly improving the utilization and efficiency of each existing fiber link, all without requiring additional construction and right-of-way costs (etc.) to lay additional fiber into the ground.

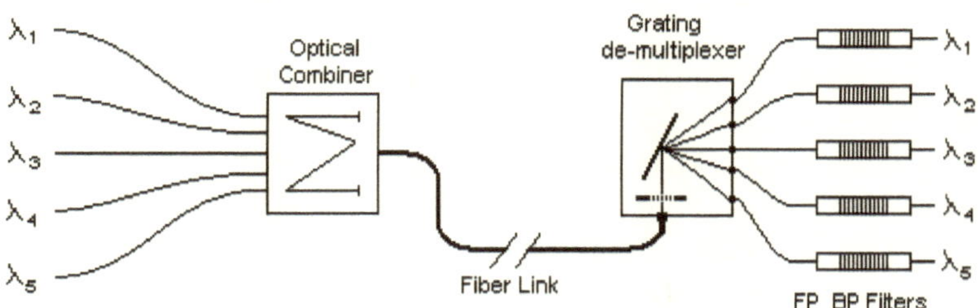

Figure 8.12 WDM telecom link.

Currently, manufacturers are able to fabricate filters/multiplexers capable of placing wavelength "channels" to within ± 0.025nm of their specified location, with wavelength channel spacing less than 1nm, Bandwidths of >0.5nm, and adjacent channel rejection of >30dB (i.e. greater than a factor of 1000)[47].

"Dense" WDM techniques coupled with Raman type fiber amplifiers currently being marketed (with gains ranging from 10 to 20 dB, specifically designed for 10 – 40 Gbs long-haul link applications) are pushing the total throughput horizon for a single fiber link to well beyond the Terabit (10^{12} bits/sec) range[48].

OTDRs:

Since no discussion about Fiber Optic Telecommunications would be complete without a respectable mention of "*Optical Time Domain Reflectometers*" (OTDRs), we include a brief overview here (see also Appendix C).

While discussing the law of Entropy in our opening remarks, we probably should have mentioned its well-established, and amply demonstrated cousin, the Law of Murphy (and its associated corollaries). Noting this glaring omission (for which we profusely apologize) we cite it now, in that it is one of the more fundamental realities of the universe on which the whole Telecommunications industry hinges (not to mention every other technology known to man).

Considering the fact that any number of events can and will occur to damage or disrupt our well-placed, well-engineered optical fiber link (cue Mr. Murphy), including the ubiquitous "Back-Hoe fade" and "dance of the sewer rat-king", not to

---

47 Where dB = "decibels" = dB = 10 log (final / initial). Therefore, a 3dB decrease in signal is equivalent to decreasing the signal by a factor of *two*; while a 6dB drop would be a signal reduction by a factor of *four* (etc.). By writing our gains or losses as dB, we change what would have been an exercise in multiplication and division into an exercise in addition or subtraction. For that reason, most Engineering texts tend to quantify things in terms of dB.

48 Lucent Technologies' Bell Labs in June 2001 estimated that due to noise, bandwidth and information density limitations (3 bits per Hertz), the maximum theoretical throughput limit of a single fiber link is in the 100 Tera bits per second range (100×10^{12} b/s).

mention flood, earthquake, fire, and general act of God, we know that sooner or later, our link (like everything else on this "mortal coil") will fail. It is only a matter of time[49].

Since a single long-haul link can now easily run over 50 miles, and considering that there are *millions* of miles of such fiber links buried throughout the US and Canada alone under virtually every city street and country highway, having a means to locate and identify the inevitable link failure quickly and easily to within less than a meter is absolutely vital if the technology is to survive and expand. After all, going through the process of digging up even one block of city street, not to mention a 50 mile stretch of countryside to locate a failure every time a link goes down, would mean certain political, social and financial death to the technology, not to mention any company foolish enough to rely on it.

Fortunately, the technology itself provides us with a very powerful solution to finding all such link failures: "Optical Time Domain Reflectometry" (OTDR).

We know from previous discussions, that anytime a wave of energy (in this case, light) traveling through a medium encounters a change in the propagation characteristics of that medium (broken, chewed, burned, cut, torn, eaten, or otherwise mangled fiber cable), some portion of that energy will be reflected back towards its source. Since the speed of propagation (i.e. the index of refraction) is a very well-known parameter for all fiber cables, we can use that information to locate exactly where the reflecting event lies simply by launching a test pulse of light down the fiber and measuring the time it takes for the pulse to travel down the fiber, reflect off the "event", and return back to us. We then simply take that total round-trip time and divide by two (to find the one-way time to the event). We then multiply that time by the velocity of propagation to find the distance to the reflecting event:

---

[49] In all seriousness, this outcome is guaranteed by the Second Law of Thermodynamics: All things left exposed to the random forces present in every environment are eventually degraded, moving everything in the universe from the complex, towards the simple.

$$\text{Distance (m)} = \text{Speed(m/s)} \times \text{Time}_{Total}\text{(s)} / 2$$

Example: OTDR reflection

If it takes 55 microseconds (55×10^{-6} seconds) for a test pulse to travel down to a reflecting event in a fiber (e.g. bad splice, broken segment, etc.) and back, how far is it to this "event", given the index of refraction of the fiber is 1.46 (at this wavelength)?

Using the fact that: $n = c/v$, we find that the speed of light in this fiber (at this wavelength) is:

$$\begin{aligned} v &= c / n \\ &= 300 \times 10^{6} \text{ m/s} / 1.46 \\ &= 205.5 \times 10^{6} \text{ m/s} \end{aligned}$$

Therefore, the distance to the reflecting event in the fiber is:

$$\begin{aligned} D &= (205.5 \times 10^{6} \text{ m/s}) \times (55 \times 10^{-6} \text{ seconds}) / 2 \\ &= 5{,}651.3 \text{ meters} \end{aligned}$$

A basic OTDR has at least one laser "transmitter" (which is used to send a pulse down the fiber), and a detector which measures all energy reflected back up the fiber. Since the reflected light is typically very weak by the time it makes its way down and back in a long fiber, the detector must be extremely sensitive. In order to prevent the very strong transmitted light from "swamping" or damaging this sensitive detector, a highly selective directional coupler, or "circulator" is used just inside the OTDR to keep each direction/path separate from the other, or the transmitted laser energy could easily damage the detector. Light being reflected back from the fiber re-enters the OTDR and is routed by the circulator into the detector, which measures both the reflected energy's intensity and time of flight (the time between when the pulse was sent and when it returned).

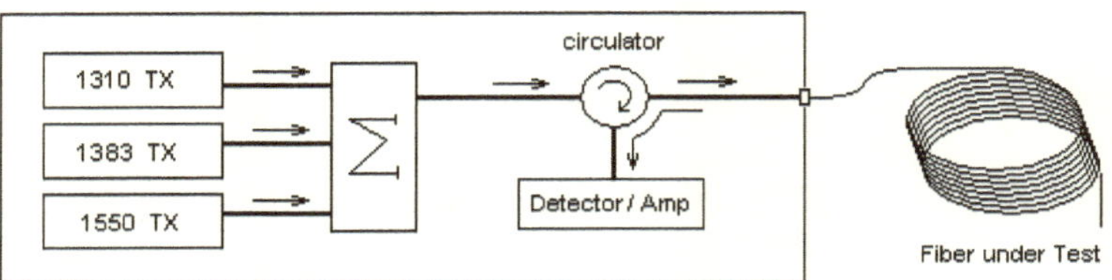

Figure 8.13 Typical OTDR block diagram showing three possible laser Transmitters.

In addition to protecting the detector from the transmit pulse, we need to insure that when we connect our OTDR to the fiber being tested, that we keep the discontinuity of the front panel connection as low as possible for the same reason. If the connection is poor (e.g. contaminated by dirt, mashed to pieces while attempting to connect two dissimilar type of connectors, etc.), a high reflection at the panel connection will result. Since this reflection at the OTDR port has experienced no attenuation through the fiber, it will be almost as strong as the energy emitted by the laser. Such a high intensity reflection so close to the detector will swamp the detector – effectively making it somewhat "snow blind" for a brief instant. In other words the detector will be driven into saturation, requiring a short "recovery time" after the bright reflection subsides, for the excess carriers in the detector to dissipate. During this time, the detector will not be able to resolve any defects in the "near" section of fiber being tested.

By using the attenuation model mentioned at the beginning of this chapter:

$$I_{(x)} = I_o \exp(-a\,x)$$

we can plot the intensity of the backscattered light throughout the length of the fiber, either as a linear plot which decays exponentially, or as a log plot, which decays at a straight line rate of "α" dB/km (see Figure 8.14).

Entropy being what it is, we know that some noise corruption will inevitably creep into our OTDR sample (e.g. "shot noise" in our electronic detector and amplifier, stray electromagnetic fields from some unrelated circuit, etc.), which could be misinterpreted as a reflection event. Since such noise corruption tends to be random (while reflections off the fiber defect tend to be consistent), if we take several such "pulse-and-listen" samples and then average them together, we can effectively average most of the random noise out of the measurement.

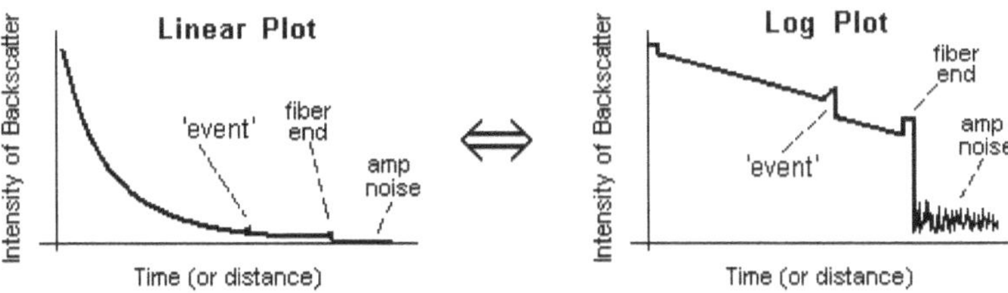

Figure 8.14 Plots of Backscattered light as measured by an OTDR, showing an event reflection (e.g. a reflection off a fiber "defect", poor splice, bad connector, etc.). Note that the width of each "event" in this trace is a function both of the event's length, as well as the test pulse being transmitted by the OTDR.

For that reason, OTDRs typically send hundreds or thousands of pulses down the fiber (with sufficient time between pulses to allow light to reflect from the farthest end of the fiber). All of the reflections from each pulse are then distributed by time across an array of software "bins" via the detection software, effectively averaging out all the random noise in the process.

The easiest way to process all of these detector samples, is to convert the analog signal exiting the detector/amplifier into a stream of digital numbers (using an "Analog-to-Digital", or "A-to-D", converter). The processing software then distributes these digital samples across counting "bins" as a function of time for each pulse, and the contents of each "bin" is then summed and divided by the total number of samples.

A real event will consistently generate reflections at the same point in each sample (and thus accumulate energy into the same counting bin), while random noise events both increase and decrease the sample energy/count per counting bin randomly. As a consequence, the energy from random noise tends to average out over time. For this reason, the Signal-to-Noise ratio ("SNR") of the OTDR sampling improves significantly as we increase the total number of samples taken and averaged.:

SNR improvement α SQRT(N)

where "N" is the total number of samples taken.

Example: SNR improvement #1

What is the improvement of SNR compared to a single sample if we average 8,000 samples?

SNR improvement α SQRT(8000)
= 89 = 19.5 dB

Example: SNR improvement #2

What is the improvement of SNR compared to a single sample if we average 32,000 samples ?

SNR improvement α SQRT(32000)
= 179 = 22.5 dB

Note that in these examples we increased the number of samples taken by a factor of *four* (from 8000 to 32000), but only improved our SNR by 3dB. This means that even though we took *four* times longer in the second example to acquire these extra samples, our Signal to Noise ratio only improved by 3dB (i.e. *double*). In other words, though acquiring more samples helps improve our OTDR measurement, there is a diminishing return for the amount of time we invest to take these extra samples. One very strong motive for sampling longer however, is that it not only improves the resolution of our OTDR measurement by averaging out the effects of noise on the sample, but in so doing, in the process it extends the range that our OTDR is able to probe down the fiber. This effectively increases the distance we are able to see clearly down the fiber in our search for defects, compared to what we are able to resolve after only a few samples.

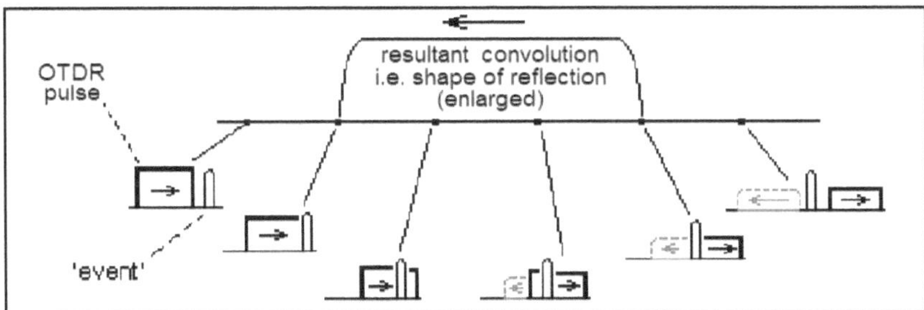

Figure 8.15 A "cartoon" of snapshots depicting the convolution sequence of a light pulse reflecting off an event. Note the total energy injected into the fiber via the OTDR pulse equals the *sum* of the energy Reflected + the energy Transmitted through the event: $E_{OTDR} = E_R + E_T$.

A second technique we can use to increase the probable distance down the fiber is to simply increase the amount of energy we dump into the fiber. Of course we could accomplish this by using a more powerful laser, however that entails extra component costs, as well as an increase in the risk of possibility of eye injury to anyone using it. Another problem we eventually encounter with more powerful lasers is that at some point we begin to induce some non-linear effects in the fiber.[50]

---

50 Recall from our discussion about the restoring force of a molecular bond and the parabolic shape of molecular energy modes, that too great of an energy injected into a molecular bond distorts the

Instead of changing out the laser in the OTDR to increase the amount of energy it injects into the fiber, a simple alternative is to make our pulses *wider*, since the total amount of energy we inject into the fiber depends not only on the amplitude of the pulse, but also on its width. This does indeed increase the distance we can resolve down the fiber (by effectively increasing our Signal-to-Noise ratio), but it also creates one slight problem:

If we wanted to characterize the reflecting event (e.g. measure its width), a wide test pulse would give us very poor resolution of the event (again by analogy, if you wanted to profile the ridges of your fingerprint, you need to use a probe that is smaller than the features you are trying to resolve). This is due to the fact that the reflection at the event is effectively a "*convolution*" of the physical width of the event, with the width of the pulse of energy we throw at it (see Appendix B).

In other words, the width of the reflection we sense at the detector depends both on how physically large the reflector is, and how long it is exposed to the energy in our test pulse. Anytime we convolve two "functions" (i.e. the event's physical width against our pulse width), the resultant product "inherits" characteristics from each of the "parent" functions. Even if the physical length of the event were extremely narrow, the reflection from our test pulse will be at least as wide as the test pulse, since all energy in the pulse (front, back and everything in between) takes time to encounter and then reflect off the full event.

Stated another way, the reflection starts when the energy in the front of our test pulse reaches the reflecting event, and does not stop until *all* of the energy in our test pulse hits and reflects off the event. Consequently, the wider the test pulse is, the wider the reflection will be off of the event.

For that reason, most OTDRs allow the operator to select pulse widths for the particular application or operation needed: *narrow pulses* for high resolution of events (e.g. factory testing of short spools where quality of fiber and defect characterization is

---

parabolic response. In the extreme cases, this will begin to rupture the bonds holding the molecule together (both being decidedly non-linear effects).

acutely important), and *wide pulses* for long range resolution (particularly useful when troubleshooting broken long-haul fiber links in the field).

Of course there are other tricks we can use to improve the "Dynamic Range" of our OTDR, including switching to a higher gain on our amplifier for the more distant (i.e. weakest) reflections, and using a "random noise generator" to insure that each sample is sufficiently "*dithered*" around adjacent A-to-D steps, etc. This latter technique helps us overcome "*Quantatization noise*", which is due to the fact that the full range of the A-to-D is broken into a limited number of very discrete steps. By injecting a *controlled* amount of randomness in our sampling process, we insure that our samples are randomly distributed across adjacent A-to-D bins. This improves our smoothing when we average the samples in each bin.

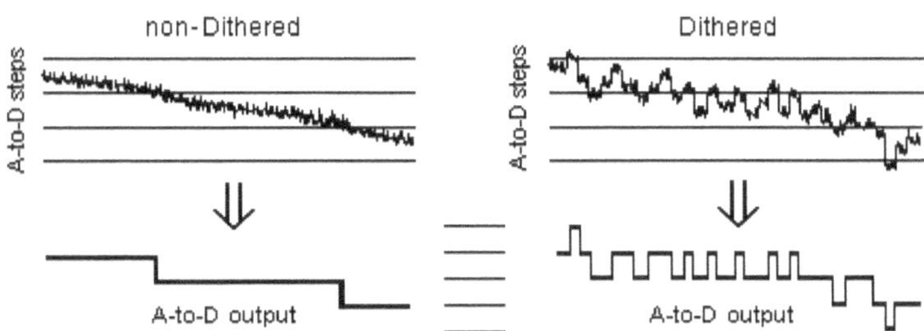

Figure 8.16 data and A-to-D Quantatization noise with and without "Dithering".

Though at first blush, it may seem a little odd to *add* noise to our signal in order to *improve* its noise response, we need to remember that a signal represented with heavy "quantatization noise" (i.e. the "stair-stepped" look) is itself a degraded signal. By dithering *each* OTDR sample slightly in a controlled fashion, we are randomly distributing that sample out across several A-to-D bin thresholds, and in so doing we reduce the quantatization effects. And since *each* OTDR sample is dithered *differently* (and therefore distributed through adjacent bins differently), when we sum and average all the OTDR samples together, the end result not only has very little quantatization noise, it also has a significant reduction in random "ambient" noise in the process as well.

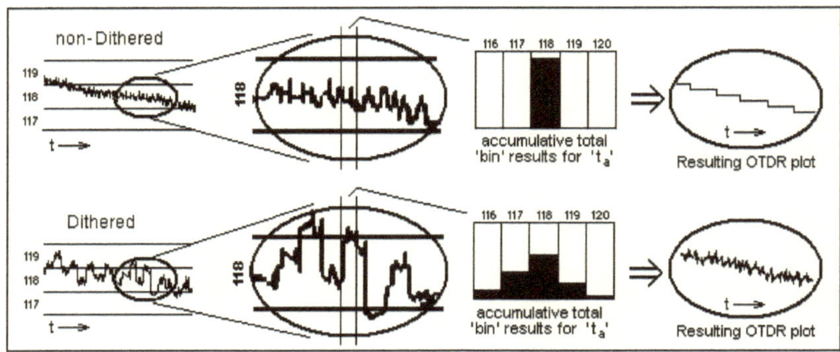

Figure 8.17 Results of dithering many "binned" samples.

OTDR Performance characteristics:

Dynamic Range – is a measure of the full range in signal strength that the OTDR is able to accurately use to resolve an event in a fiber. In other words, the total span of usable signal strengths from the strongest reflections (closest to the OTDR) to the weakest of reflections (farthest distance down the fiber) that an OTDR is able to use to accurately resolve "defects" in the fiber link. Typical values achieved by current OTDRs on the market are around 35-38 dB, depending on wavelength (3 min averaging, 1000 m pulse width). Since fiber attenuations are now typically around 0.35 dB/km (for a given wavelength), this implies a maximum probing distance of ~100km down a given fiber.

Linearity – is a measure of how "consistent" the response of the OTDR is over its range of measurements. Since the absorption of energy in a fiber tends to produce reflections that are relatively weak, all OTDRs employ some form of an amplifier immediately following the detector. The *ideal* amplifier will apply a *uniform* gain factor to *all* signals regardless of signal strength. In practice however, no amplifier can maintain a perfectly uniform gain for all signals across the entire operating range of the OTDR. As a result, this introduces a slight non-linearity in the response of the OTDR (e.g. stronger reflections are amplified slightly more than weaker signals, or signals that fall within a narrow range of intensities, etc.).

To test for non-linearity, two samples can be taken of a long fiber in quick succession (using the same pulse width, wavelength and averaging time) and inserting a 1-2 dB attenuation in line during one of the samples. These two samples are then subtracted from each other. Since the fiber is identical in both cases, the *ideal* profiles of the fiber should be identical for both samples, and when the two are subtracted, the result should be a flat line (in the ideal case). However in practice, the non-linear effects in the amplifier response typically cause the two samples to be slightly different. This difference becomes immediately clear when the two samples are subtracted (i.e. the resulting difference is typically not a perfectly flat line). Most OTDRs on the market today typically have linearities of less than 0.03 dB/dB (i.e. their worst deviation from zero in the difference between the two samples is less than 0.03 dB of difference per dB of signal strength).

Distance accuracy – is a measure of how well the OTDR can locate defects along the full length of a given fiber. Since the primary purpose of an OTDR is to be able to accurately resolve and locate a "defect" in a fiber, it is very important that the measured distance value be as accurate as possible. However, as mentioned previously, as we increase the width of our pulse we send down the fiber from the OTDR lasers, we decrease the ability of the device to accurately resolve the reflecting event. In addition to this effect, we also note that since light travels extremely fast (300×10^6 m/s), resolving the true location of an event implies being able to resolve reflected energy down to the nanosecond (i.e. $\sim 10^{-9}$ seconds). Once the reflection is sensed at the detector, it goes through a series of processing steps, including amplification, conversion into a digital representation by the A-to-D, routed into a memory location (i.e. "binning") as a function of the amount of time lapsed after the pulse was injected into the fiber, and an assortment of possible averaging, smoothing and filtering operations (which varies by manufacturer), all typically performed by some digital processor (CPU or DSP) .

All of these processes not only require time, they also introduce some finite amount of uncertainty in the amount of time they contribute to the overall processing phase of the analysis. Consequently, there is always some element of uncertainty in the time measurement of each reflection, which translates into an uncertainty or ambiguity as to the true location of the event as measured by the OTDR. Most OTDRs on the market today typically exhibit less than 0.05m location error using a 5m pulse width, and less than 0.25m error using a 100 meter pulse width.

Spatial Resolution – Is a measure of the OTDR's ability to accurately resolve and measure the size of all reflecting events, without introducing any distortions due to its own imperfections. The event portrayed by the OTDR display is produced by amplifying and processing the reflected energy from the event. Since no amplifier has perfect, instantaneous response, and since the conversion and manipulation of an analog signal into a digital representation inherently implies something less than a faithful and perfect processing (e.g. the introduction of quantatization noise, etc.), we expect our OTDR to distort the width of the event to some degree or another. The "Spatial Resolution" characterization of the OTDR quantifies this indirectly by specifying the measured width of a known event, using the standard test configuration shown in Figure 8.18.

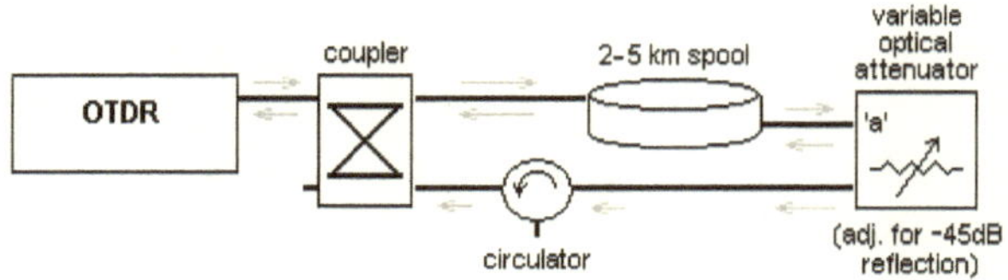

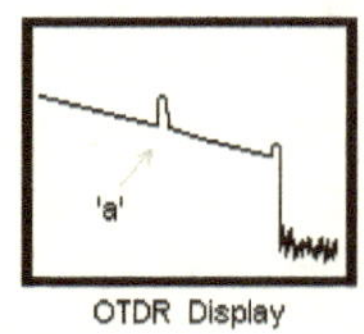

Figure 8.18 Spatial Resolution Test Fixture, SRTF. The Transmitted pulse travels out from the OTDR, through to the end of the delay spool, and back via the circulator (shown on the OTDR trace as the line segment up to point "a" on the OTDR trace). A second path also exists, which is a reflection off the attenuator and back through the delay spool a second time and then back to the OTDR (the line segment following point "a").

Dead Zone – is a measure of how quickly the OTDR's detector comes out of saturation-induced desensitization following a bright reflection. When a reflection arrives back at the OTDR detector[51], the photons from the reflection stimulate the release of electrons out of the detector via the photoelectric effect. Large discontinuities in the fiber produce very bright reflections, which in turn stimulate a large avalanche of photons. In extreme cases this tends to create something of saturation effect in the avalanche diode. During such a saturation a very large number of electrons accumulate in the detector, which requires some period of time to drain well after the photons reflected off the event are gone. When the diode is in saturation, its response to any other reflected energy is degraded, in some cases rather severely.

As a result, there is typically a brief interval following large reflections in which the detector is desensitized. The severity of this desense interval is a function of the intensity of the reflection (the more intense the reflection, the wider the "blind spot" – a common example of this problem occurs when the

---

51 One of the more common type of detector used is the Avalanche Photo Diode (APD). Typically, a high bias Electric field is applied across the APD which accelerates the electrons ejected by incoming photons. As these electrons travel through the diode, they collide with other electrons which then in turn are also accelerated by this external field and similarly generate even more electrons (etc.) in an avalanche release of electrons. This overall process effectively allows a single photon to generate hundreds of electrons, forming an output current pulse from the device, which is then subsequently amplified and processed.

OTDR-test fiber connection is poor). Obviously, it is important to be able to quantify how quickly the detector recovers from such a saturation. This can be indirectly quantified using the same test fixture shown in Figure 8.18 by measuring the width of a known event, with particular attention paid to the trailing edge of the event.

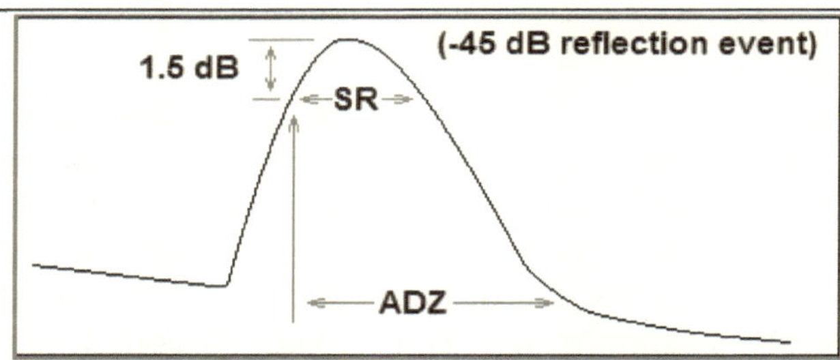

Figure 8.19 "Spatial Resolution" and "Attenuation Dead Zone"

Spectral Purity – is a measure of the OTDR's laser spectral profile. From our previous discussion on chromatic dispersion it is clear that whatever laser we use as our transmitter in the OTDR, its spectral profile is critical. If the output of that laser has too broad a range of wavelengths, Chromatic Dispersion would tend to degrade our ability to accurately locate and profile any defects along our fiber length (see Figure 8.5). This characteristic is typically quantified using an optical Spectrum Analyzer (OSA) to insure that the spectral profile of the laser is sufficiently narrow (e.g. a Full Width, Half Maximum of < 2 – 5 nm).

Research and Development:

The following section is meant as a brief overview of a few current areas of research intended to provide future improvements to laser applications, and fiber optic telecommunications in particular.

Soliton pulses:

In Chapter 4 we briefly introduced soliton waves as a classical analogue to the traveling electron wave-packet. We pointed out that solitons waves have been studied scientifically since the early 1800's due to their unique ability to travel great distances while losing almost no energy. As it turns out, part of the secret to soliton's ability to suffer minimal energy loss as they travel is related to "chromatic dispersion".

In our previous discussions we described chromatic dispersion as an effect in which different frequencies travel through a medium at different velocities, separating the frequencies as they travel (a common example being the creation of a rainbow). To put a finer point on this effect, we note that pulses of energy (e.g. sound, E&M waves, sharp impulses injected into metal objects, etc.) represent a *composite* of many frequencies, as demonstrated in our initial discussion on superposition (see Figures 3.5 and 3.6). Furthermore, from studies on chromatic dispersion, we know that the higher frequencies (sometimes referred to as "*Brillouin Precursors*" in E&M texts [4]) typically tend to propagate *faster* than lower frequencies.

The key to soliton creation, is in the fact that under certain conditions, this speed-related *frequency* separation can be exactly offset by a competing *amplitude* mechanism that affects the speed in exactly the opposite way. This amplitude-dependent effect causes higher frequency components (with the right intensity) to "linger" a bit longer in the media compared to lower frequencies, effectively slowing them down slightly.

Though the index of refraction ($n_{(F)} = c/v$) is definitely frequency dependent, we find it is also affected by the *intensity* of the light in a non-linear way. In fact, the brighter the light is, the *slower* it is propagated by the atoms within the media (this is analogous to the swamping of our photodiode in the previous section, in that very bright light saturates the detector, requiring a longer time for that stored energy to dissipate once the incident light is gone). One of the consequences of this effect is self-phase "modulation", where the more intense frequency components begin to *lag* the weaker components as they propagate.

Under certain conditions these two competing mechanisms can exactly counteract one another, producing Solitons. As a result, under these conditions the energy that the pulse starts with will dissipate very slowly, allowing the pulse to propagate extraordinarily long distances by comparison to normal wave propagation.

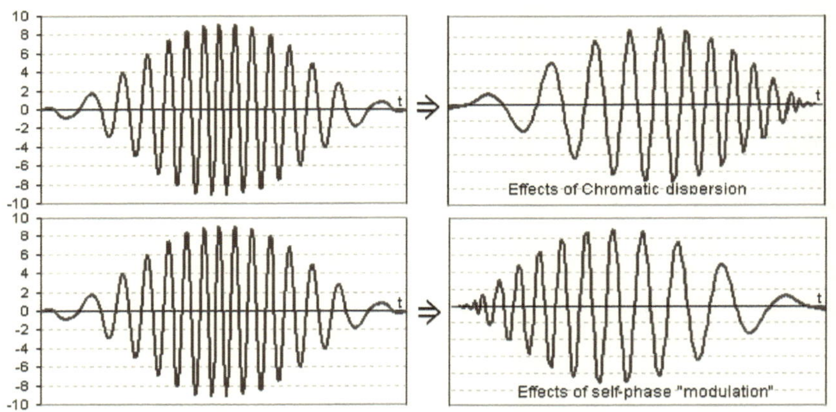

Figure 8.20 Effects of Chromatic dispersion and non-linear Self-phase "modulation".

In applications where we need to send wave packets across large distances (e.g. over long-haul fiber optic links), having a pulse that travels great distances with very little energy loss is a very appealing concept. Unfortunately, all fibers unavoidably induce some amount of scattering and absorption of the energy we inject into them, be they our "standard" light pulses, or solitons. As a result, even solitons eventually begin to lose energy, causing them to not only diminish in amplitude, but also spread as they travel. However, their ability to persist much longer than normal pulses of light, offers the promise of a significant extension to our current maximum long-haul segment lengths (currently at ~100 km, or ~65 miles). In the mid 1980's,

Bell Labs reported that they were able to successfully use solitons in a fiber optic link, claiming some persisted for an impressive 6000 km with minimal dispersion and loss.

Photonic Logic Processors:

During a long-haul transmission from point "A" to point "B", we typically convert the signal from photonic, to electronic, and back again multiple times. As we found in our previous discussion on Fiber Amps, there are strong motives for avoiding the down-converting of photons into electrons and back again. However, at the present all of our logic processing systems still depend heavily on multiple up/down-conversions. The reason is simple: the conversion of our signal into electronic form allows us to not only amplify our signal, but also perform the kind of logical operations on the signal that we currently are unable to duplicate in the photonic realm.

One of the simplest examples of this logic processing is the common "AND", "NAND", and "OR" gates, forming *the* quintessential building blocks of all modern computers. With these basic gates, a computer processor is able to simulate the fundamental aspect of reason, and thus perform basic logical decision based on constantly changing inputs.

The behavior of the "AND" gate is such that the output goes "high" (i.e. "true", or a logic "one") *only* if *both* input signals are in the "high" state. Otherwise the output of the AND gate remains "low" (i.e. "false", or a logic "zero"). This basic logic gate provides us a very simple and yet very effective way to enable a machine to decide an outcome based on a simple set of conditions present at any instant in time.

In addition to the AND gate, there are also "OR" gates, "NOT" gates, etc., which are also important when constructing a simple electronic decision processor (the schematic symbols for these gates and their basic "truth" tables are shown in Figure 8.21).

By connecting several of these gates together in a specific way, we are able to construct an electronic processor with a well-defined logical decision making process. Such a logic circuit can take several input signals and use them to activate a specific output or sequence, depending on the state of each of the inputs at the time (e.g. the classic "cocoa vending machine" example, in which the right coins must be deposited, AND cocoa must be available, AND a cup dropped into the dispensing chute, AND water must be above the minimum temperature, but NOT above a maximum temperature, before hot cocoa is dispensed into the cup slot, etc.).

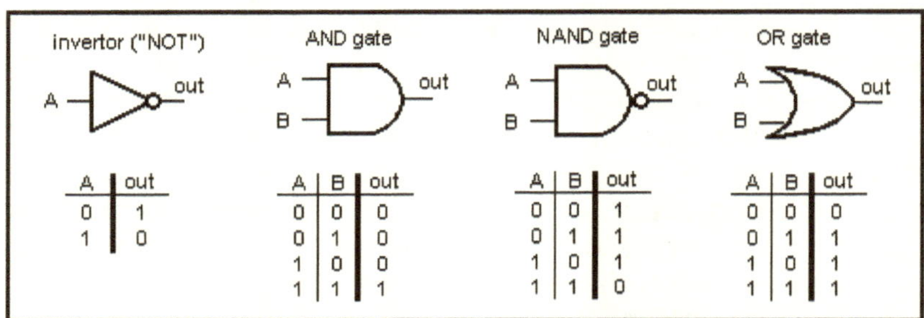

Figure 8.21 Basic NOT, AND, NAND, and OR logic gates.

Since photons can travel much faster than electrons in a circuit, there is a significant speed advantage in being able to develop an array of logic processors that works entirely at the photon level (not to mention all the noise immunity advantages, etc. mentioned earlier in this chapter). If optical equivalents of these logic gates can be developed, the semiconductor fabrication techniques described in Chapter 5 and the advancements of Large Scale Integration (LSI) semiconductor fabrication techniques that helped make the desktop computer ubiquitous, could similarly be applied to the optical versions of these semiconductor devices, jump-starting the development of any possible optical processor.

As the dimensions of the *electronic* versions of these components fabricated into semiconductors has continuously *decreased*, the speed at which each electronic gate is able to process data has constantly *increased* (enabling CPU processing now measured easily in the hundreds of Millions of Instructions Per Second, or MIPS).

Obviously the techniques used in fabricating such LSI devices offers us an incredibly powerful tool in the manipulation and routing of electronic signals. The electronic gates however have two significant problems or limitations:

1) At the heart of how these electronic circuits work is the humble little electronic charge. From Chapter 6, we know that anytime a charge moves, it radiates an electromagnetic field (Ampre's Law). Conversely, anytime an external field approaches a charge, it tends to displace that charge (or group of charges; Faraday's Law) by applying a Lorentz force on it. As a result, electrons moving through one circuit tend to create fields that often "couple" into adjacent circuits and inadvertently induce an undesirable flow of current in the adjacent circuit (i.e. the currents in one circuit often "bleed" into adjacent circuits), creating what is often referred to as "cross-talk".

In addition to "cross-talk", we know that other extraneous E&M fields (i.e. noise) from external sources often couple into our electronic circuit, whether that is produced by an adjacent circuit, a flash of lightning (representing a *massive* movement of charges and thus *huge* fields), or anything in between. In fact, there is a seemingly infinite supply of possible noise sources all around us, ranging from anthropomorphic sources (e.g. leaky spark plug wires in cars, vacuum cleaner motor brushes, high power transmission lines and transformer substations, etc.) to natural environmental sources (e.g. the continuous barrage of global lightning-induced noise, solar flare ion storms, quasars/black holes/time-traveling Deloreans, etc.). All of this noise naturally tends to find its way into our humble little electronic circuit, degrading the order and clarity of our signal in the process (again, as predicted by the Law of Entropy).

2) The expansion of electronic technology over the past five decades faces another type of problem, that is related to what is known as "*Moore's Law*": Moore's law (named after Gordon Moore, who coined it in 1965), states that the smallest size of the features that can be fabricated in semiconductor *halves* approximately every 18 months, which implies that the number of components that can be fabricated in semiconductor

doubles approximately every 18 months[52]. By Moore's law, within the next two decades, dimensions will become so small as to reduce gate dimensions and paths to a few tens of atoms wide.

Currently, dimensions of gates and wires fabricated in semiconductor ICs are large enough that current flows are on the order of ~10^{15+} electrons per second and greater. This vast number is large enough to allow us to ignore quantum effects (the Uncertainty principle, and tunneling in particular). However, if Moore's law holds through the next 20 years, dimensions will become so small as to result in currents of perhaps only a few hundred electrons per second. At that point we will have encountered the "Quantum threshold", where a few electrons going awry out of a few hundred is no longer trivial, thereby making such Quantum effects as Uncertainty and tunneling very significant. This will require us to migrate from the relatively simple rules we have been accustomed to using in electronics such as Ohm's Law and Kirchoff's rule, to using much more complex techniques centered around Quantum Physics and Schrödinger's equation to describe and predict the behavior of these sub-microscopic circuits.

In addition, as we approach this limiting threshold we will find it much harder to pack ever greater numbers of transistors into the same piece of silicon, or germanium (or diamond, etc.), due to the resolution possible using visible wavelengths of light. There is the possibility of trying to use ever-shorter wavelengths (e.g. x-rays), however focusing shorter wavelengths is much more difficult due to the index of refraction approaching one at these frequencies. This will make it that much harder to increase the speed and complexity of our signal processors in order to keep pace with the growing demand.

Consequently, these complications represents something of a soft upper bound for the advancement of current semiconductor techniques and all the technologies designed around it, including fiber optic processing systems. In the past as we have reached the limits of an existing technology (e.g. the

52 Current minimum feature resolution and fabrication dimensions is approximately 0.18μm.

processing limit of mechanical gear-driven "Analytical Machines", the limits of computing via analog circuitry, the limits of relay/vacuum tube signal processing, the throughput limits of copper phone lines, etc.), the solution was to retire that technology by developing something completely different with a higher upper bound. Since the demand for greater data throughput can only be expected to continue to increase through the foreseeable future, some alternate approach will be needed to accommodate this ever-expanding demand for improved speed, capacity, and bandwidth.

To that end, a great deal of effort has been focused on attempting to make the next "quantum" step in technology, by developing equivalent photonic processing systems akin to the now familiar NOT, AND, NAND, and OR logic gates. Since photons carry no charge and no mass, they are not subject to the same limitations that degrade systems designed around the flow of electrons, including the degrading influence of external noise fields. In addition, being charge-less, they do not create fields when they move, and hence do not develop reactances that impede their flow (as is the case when electrons move through a circuit, creating inductive and capacitive reactances). And since these reactances are significant factors restricting bandwidth, any system not hampered by them offers the promise of much wider bandwidths and hence greater data throughput than their equivalent electronic counterparts – not to mention near *light speed* processing.

However, the lack of a charge on the photon is in the same instant both a blessing and a curse, since it leaves our current technology without the "standard mechanism" to enable two signals to interact to effect a logical decision on a strictly photonic level. From our earlier discussions on our linear Maxwell's equations and the fact that fields/photons mix only in non-linear media, we find our typical optical media offers us no clear way make two photon streams interact in such a way as to facilitate a photonic logic gate. This suggests that a solution might be be possible if we can control and implement a non-linear Quantum process, such as Raman Scattering (etc. see below). Should a viable solution be discovered that could create a working photonic logic processor, its discovery would revolutionize not only computing and fiber optic telecommunications, but in the process have a profound impact on virtually everything else we do.

To illustrate some of the challenges involved in creating a photonic gate, consider the problems in the following simple approach to creating an optical "AND" gate:

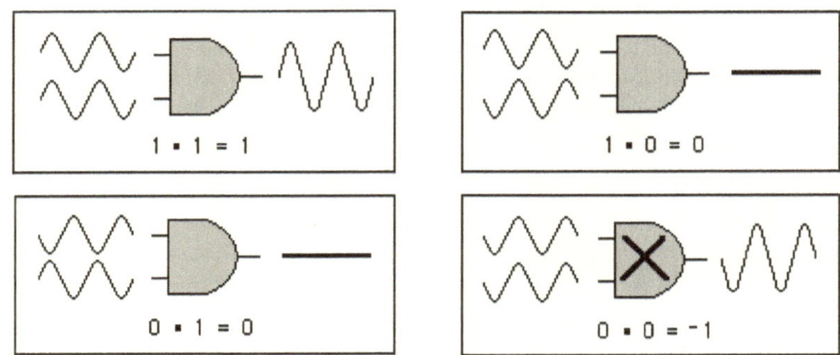

Figure 8.22 An optical "AND" gate based strictly on phase. Note in this simplistic approach, only three of the four "AND" conditions can be simulated, leaving the fourth unrealized. Further, any solution based on phase faces significant complications due to temperature, humidity, (etc.) variations. For any optical solution to be viable, it must provide consistent performance with minimal degradation due to external variables such as temperature.

If we were to exploit phase-related constructive and destructive interference as the fundamental method of interaction between two signals, we find that when both input signals are in-phase, they superimpose constructively, generating a "high" output. When they are out of phase, they destructively interfere, generating a "low" output. Using this approach, we find that we can complete *most* of the conditions for the full truth table for an AND gate, but typically find it difficult to realize *all four* conditions simultaneously.

In addition, since the performance of such an approach depends strictly on the phasing of each input, and since phase itself is heavily impacted by wavelength, temperature variation affects on physical dimensions, phase delays, etc., the success or failure of such an approach is critically dependent on placement, temperature induced path variations, and a whole host of other variables, further complicating its use (see Figure 8.22).

A more viable approach would be to eliminate the phase sensitivity entirely by relying on the *intensities* of the input signals, and then routing each input signal into a non-linear device that responds to intensity only. Borrowing once again from our previous discussions about Maxwell's equations and linear media, we know that the only time two signals will mix is when they are traveling through some non-linear device, such as a diode (or transistor).

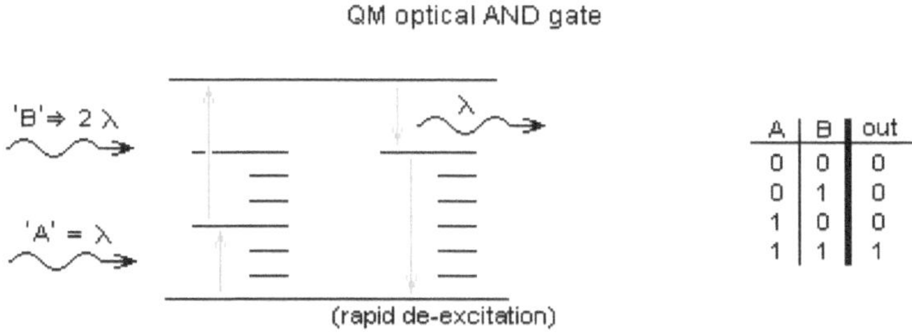

Figure 8.23 Simplified version of a possible QM optical "AND" gate based on a Raman-type energy absorption.

We found during our discussions in Quantum Physics, that by clever choice of molecules, we can promote non-linear interactions via Quantum behavior such as the Raman Effect. These unique cases then, suggest one possible solution to creating a simple, yet functional optical "AND" gate, by providing us a means to promote interactions between two light signals via a non-linear medium.

For example, if we could design an atomic (or molecular) system that would step to progressively higher energy states when each input signal is injected, and would then quickly de-excite by expelling a specific wavelength photon only *if and when* a specific target state is reached, we might be able to use that to create a functional "AND" gate. Of course we would have to shift the frequency of one of the inputs to achieve this stepping operation, and the atomic system would have to have well-chosen lifetimes for each state: Too short a lifetime and the atom would de-excite before the other signal arrived. We further note that though this approach is wavelength dependent, it is not phase dependent. And finally, we point out that this approach is easily extendable to accommodate "n" input signals, simply by developing additional quantum levels to serve as the additional "ladder steps".

Another approach would be to use the electric field of one signal to align the molecules in a dielectric compound, and thereby cause that compound to act as a polarization filter. The degree to which the molecules are realigned (and hence the effectiveness of the polarization filter) depends on the strength of the electric field (or intensity). Such a switchable polarization filter suggests an effective way to construct a gate which can use photons from one source to alter the path of a second signal. A common example of such a device is known as a "Kerr cell", something that has been used for quite some time to form output shutters on laser cavities, which allows a particularly effective way of pulsing the output of a laser in a very well-controlled and precise way.

One of the biggest problems most of these techniques face, is the sensitivity they tend to have to external variables such as heat, timing, stability, etc. In a sense, many of these sensitivity issues almost draw parallels with old analog logic circuits, which were notoriously "finicky" when it came to variation with temperature, humidity, etc., and which ultimately made analog computers unreliable (prompting the advent of digital electronics).

A great deal of research is currently being devoted to finding an ideal solution for optical logic processing. If a viable optical solution can be found, it would not only offer huge benefits to fiber optic telecommunications, it would completely revolutionize computing itself in a way that would be even more significant than what occurred with the advent of the desktop digital computer, smart phone, etc.

Chapter 8 Review questions:

1. What is "bandwidth" and "energy spectral density" and how do they affect the flow of information?
2. What are some of the factors driving the need for more "bandwidth"?
3. What happens if we try to force a spectrally rich signal (e.g. real-time streaming video) through a system with too narrow a bandpass (e.g. 3 kHz phone line)?
4. What is "high-frequency roll-off" and how does it affect a digital signal?
5. What advantages does a fiber optic link offer over old "copper" wires (i.e. how does modulating light enable us to increase link throughput)?
6. How does Rayleigh Scattering and Water absorption affect signals in a fiber?
7. What is fiber Backscatter, and what causes it?
8. Explain Total Internal Reflection and its importance in fiber optics.
9. Explain how the outer cladding layer improves light propagation in fiber optics.
10. What is Inter Symbol Interference (ISI)?
11. Explain Chromatic Dispersion. How does it limit our data throughput?
12. What are "modes" in a fiber waveguide, what causes them, and how do they impact long-haul fiber links?
13. What are Dispersion Shifted Fibers, how do they work, and how do they improve bandwidth?
14. How does a wide beam divergence from a laser degrade throughput? How do we combat it?
15. Explain the difference between Single mode and Multi-mode fiber operation?
16. Explain the Raman Effect.
17. What are the "Stokes" and "Anti-Stokes" emission, and what are their relationships to Rayleigh "scattered" emissions?

18. Why are Stokes emissions more intense (i.e. brighter) than Anti-Stokes emissions (hint: think in terms of the *probability* of each).
19. Explain how rare earth and Raman fiber amplifiers work. What advantage do Raman amps offer?
20. Explain Wavelength Division Multiplexing (WDM) and why it is useful.
21. Explain how an OTDR works.
22. Explain how sampling time, and pulse widths affect the accuracy and range of OTDR measurement?
23. Describe several ways to increase the range of an OTDR.
24. What is a soliton? Why is there interest in using solitons in fiber optics?
25. Explain how a soliton is created.
26. What is photonic processing, and why is it desirable?
27. What difference between electrons and photons make photonic processing ideal?
28. What aspects of photons make it particularly difficult to create a photonic AND or OR logic gate?
29. What is Moore's law, and what does it imply about enhancements in current semiconductor based technologies?
30. How will this impact fiber optic communications and computing in general?
31. Explain some of the challenges in creating an optical logic gate.
32. Describe a few of the possible solutions to creating an optical "AND" gate.
33. What is a Kerr Cell, and how might something similar provide a solution to optical processing?

Closing thoughts:

In the past two and a half centuries, mankind has seen a series of mini technological "revolutions" that have had a profound effect upon both our overall knowledge base, as well as our collective social and cultural realities. Examples include the development and expansion of commerce along river and canal systems, the telegraph, the steam engine and intercontinental railways, the electrical power grid, the telephone grid, wireless broadcasting, the transistor, the semiconductor integrated circuit, and the ARPAnet / internet. In a sense, each of these watershed events can almost be viewed as "perturbations" on an overall historical plot of the "industrial revolution".

Each time one of these events took place, it was almost as if a mighty hand had reached down and engaged a new gear in a global "clockwork mechanism". As with any such mechanical shift or system perturbation, there has always been something of a transient spike in activity which has caused the "system" it was in to "shudder" for a moment. Eventually such shuddering effects tend to dampen over time, establishing a new equilibrium state along our historical track that is markedly different from its previous trend.

This perturbation transient has several names and faces, but the effect tends to be nearly identical each time it occurs. In the parlance of the old west, it was known as a "boom and bust" cycle. Upon the rising edge of one of these events (e.g. the Erie Canal, or '49 gold rush, the transcontinental railroad, the US interstate system, etc.) there has always been a period of frenzied growth and rapid development as individuals, governments, and corporations frantically attempt to take advantage of the initial "swelling wave" of the "perturbing" event, followed sometime thereafter by a saturation and adjustment period from which a new equilibrium eventually emerged.

In the past, these events tended to take years if not decades to unfold, allowing more than sufficient time for the overall "clockwork mechanism" of individual communities / societies to adjust themselves and refine whatever variables were in-play. However, as these events reshaped societies and humanity in general, they have also tended to induce an acceleration in the pace at which they themselves precipitated subsequent perturbations.

For example, the deployment of steam locomotives and railways necessitated the development of ever-larger iron and glass foundries, along with the equipment and supplies to operate, sustain and expand them. From this newly developed manufacturing base grew the capabilities and needs that promoted expanded factories and assembly lines that gave birth to the trillion dollar automobile, transportation and interstate commerce "industries". These in turn fostered growth that necessitated the development of such things as the electrical power generation and distribution networks, and mass communications to facilitate their attendant development and growth, etc. From the expertise and needs surrounding the communications and electrical power industry grew the techniques and industries centered around electronic subsystems, semiconductors, etc. Each of these events, and countless others around them, began themselves to necessitate and spawn multiple other events, triggering their own changes and perturbations in this rapidly accelerating trend.

Each step along the way, older technologies and industries in the same instant both precipitated the birth of their progeny, as well as offered some element of opposition to the changes they triggered, as both wealth and influence ebbed and flowed in the process.

Recently however, such events have begun to occur in such rapid succession and with ever-widening reach, that many of the variables involved have been hard-pressed to establish any perception of stabilizing into a new equilibrium before the next wave of the perturbation / "boom and bust" cycle landed. In fact, it almost seems as though these *upheavals* have come of late in such an extremely rapid pace and of such excessive magnitude, that our overall "clockwork mechanism" is now almost in a state of constant fibrillation, often seeming on the verge of tearing itself apart.

In just the past 25 years, we have watched numerous global "constants" transform, crumble or evaporate now almost in a twinkling of an eye. Examples include the fall of the Soviet empire, the "dot-com" phenomenon on Wall Street, and the near wholesale collapse of entire economic "hemispheres" in Taiwan, Korea, Malaysia, Bolivia, Argentina, Russia, etc., all triggered in part or in whole by little more than the accelerated rate of change in the distribution of information and ideas.

Three hundred years ago, it took perhaps half a year to project a nation's influence from one hemisphere into another. With the advent of the steam engine that time span was compressed to weeks, while jet transport further compressed that to hours. Now with the advent of satellite and fiber optic telecommunications, one's influence can exist virtually, in near *real-time*, in an almost *unlimited* number of locations around the globe *simultaneously*. (e.g. the collapse of the previously mentioned economies through instantaneous global currency manipulations in the 1990's, not to mention the alarming growth of cyber warfare and espionage, etc.).

As with all other such technological revolutions, the birth of fiber optic telecommunication technology has undergone its share of transient perturbations, and in fact is still in the process of searching for its new equilibrium, as multiple companies simultaneously attempt to expand and contract various roles and needs surrounding this technology. Estimations of internet use and associated bandwidth needs are no sooner projected than they are immediately revised as numerous new social trends, events and economic pressures superimpose in a dynamic and near chaotic way (e.g. "social media"; the shift of personal computing from the desktop, to the laptop, to the palmtop/"smartphone", etc.).

Several poignant examples of this recent roller-coaster ride include the near overnight "boom and bust" cycle of such mega-corporations as Enron, WorldCom, Global Crossing, Lucent Technologies, Nortel and the internet / media super-giant AOL-TimeWarner, all of which went from reporting incredible profits around the 2000 fiscal year, to equally phenomenal losses less than two years later (e.g. $20 billion loss at Nortel and $54 billion loss at AOL-TW), not to mention the "spore-like" growth and demise of a plethora of dot-com "virtual corporations" along the way.

Such incredible swings have come in part due to the now instantaneous flow of *vast* amounts of information, driven as much as facilitated by mega-*terra*byte fiber/satellite links across the world – even as it presses for ever-greater demand for higher bandwidth capabilities in the intangible virtual realm of electronic cyberspace. Many of the factors driving these changes were themselves heavily influenced by numerous other events and perturbations, including the "dot-com" bubble of the late 1990's, the housing bubble, the post 911 economic downturn, the 2008 Global Financial Implosion, and the subsequent *emaciated* economic environment these latter events precipitated, etc.

As we brace for the next "boom/bust" bubble cycle that seems almost certain to be every bit as extreme as the past few such "boom/bust" cycles, if not worse (e.g. the *staggering* US Federal debt bubble, etc.), the best we can hope for in this now frenzied white-knuckle ride environment, is that enough reason will prevail to help these undulations settle into a new and healthier equilibrium over time.

Obviously, the question remains open as to just *where* this equilibrium point will be, and what will have to happen before reason will prevail enough to help it begin to emerge. However, since information and its corresponding spectral density are now as much a part of the ecosystem as is food and water, it seems a solid communications infrastructure will continue to be an essential part of the "foreseeable" future, whatever shape that may take.

Appendix A: Resonance

The phenomenon of resonance is actually a fairly common thing. It occurs in a wide variety of different forms, including the resonant vibrations in a tight violin string, a hallow pipe, a radio antenna, a molecule, etc. In fact, most systems can be made to resonate if sufficient energy is injected into them and internal absorption mechanisms ("dampeners") do not dissipate too much of the injected energy.

As this energy radiates through the object/medium from one end to the other, it propagates with a characteristic speed through that medium. This characteristic speed of propagation is a function of the properties of the material from which the object is made (its density, its rigidity, its elasticity, etc.). For example, anytime we pluck a tight string, we are effectively transferring energy into that string in the form of a quick impulse, which itself has no unique, single frequency to it. In a sense, the energy in such pulses can be viewed as containing a wide range of frequencies (i.e. they contain a very large number of frequency "components", as suggested by our delta pulse transform example in Appendix B). As this energy begins to propagate down the string, it does so at a speed that is effectively imposed by the material of the string (for example the vibrations injected into the "cold steel" of a guitar string propagates at a different speed than does the vibrations in a taught rubber band), etc.

When this propagating energy encounters a change in the characteristics of the string (or whatever medium the energy is propagating through), this change in characteristics makes it difficult for the energy to couple into the new region. As a result, some portion of that energy is reflected back in the direction from which it came (as a consequence of the Law of Conservation of Energy). This is most pronounced as the vibration reaches the end of the string where the propagation characteristics are dramatically different, causing the vast majority of this energy to reflect back. Reflections do not occur only at the ends, but in fact occur anytime the propagating energy encounters any significant change in the characteristics of the medium, such as a defect, a knot, or an abrupt change in size, shape or composition of the media.

As this energy then propagates back and forth between these reflecting elements in the medium, such as the ends of the plucked string, we find that a unique event develops in which the characteristics of the medium itself effectively *selects* certain frequencies out of the impulse of energy we injected when we plucked the string, "resonating" at those "special" frequencies.

This selective resonance develops due to the fact that it takes a specific amount of time for the energy wave to propagate down to one end of the media and reflect back. The longer the object is (and/or the slower energy propagates in it), the longer it takes for the wave to reach the end and bounce back. If the energy reflected from one end arrives in the middle *in-phase* with the energy reflected from the opposite end, the two reflections reinforce each other. If they arrive *out of phase*, they cancel each other out.

As a result, the length of the object and its propagating characteristics effectively determine how long it will take for the reflections to travel out and back, which inherently implies that the length effectively selects only certain wavelengths (or frequencies) as these *constructively* interfere with each other, while all other frequencies cancel each other out (as they *destructively* interfere). If we shorten the string or pinch it down against a hard fret, the wave energy has to cover less distance, and hence it is reflected back to the center sooner. As a result, the shorter string resonates at a shorter wavelength (i.e. higher frequency).

This same effect can be demonstrated in a bar of steel, a crystal goblet, or half-filled pop bottle, etc. In each case, some energy is injected into the system in the form of a hammer strike on the steel, rubbing a finger across the edge of the crystal, or pulses of air blown across the lip of the bottle (etc.). The reflecting back and forth of the injected energy between the boundaries of the object then develops resonant vibrations within this system. As we shorten the steel bar, or the empty space in the goblet, or pop bottle (etc.), we alter the dimensions of the resonating structure, and thereby alter the resonant frequency of the structure.

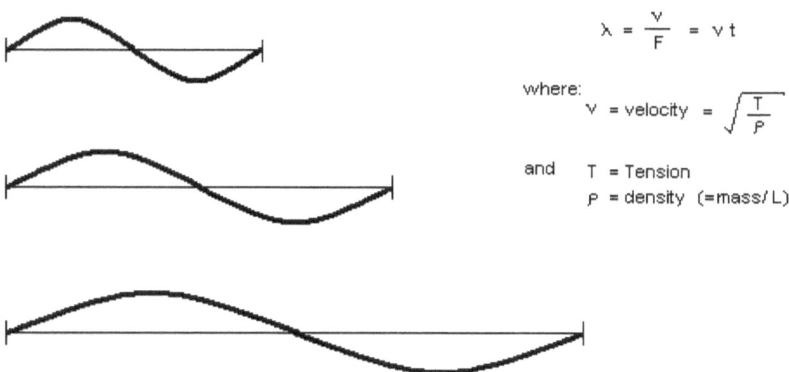

Figure A.1 Resonant wavelengths on a string of variable lengths

As it turns out, in order for these reflections to arrive back at the center in phase, the total length of the media must be some multiple of a half wavelength. To understand this, consider a sequence of waves starting from some point on the string, such as the left most edge in our figure. If the string is equal to a half wavelength of the injected energy, as one wave travels down the string and reflects back, it has covered twice the distance of the string, or one full wavelength. When it arrives back at the start, it is *in phase* with the next wavefront being injected and as a result, the energy in each successive wave is reinforced by superimposing in-phase with the energy from the previous wavefront.

All signals with different wavelengths that meet *out of phase* (e.g. as one is reaching a positive peak and the other is hitting its negative trough, etc.) cancel each other out. If the wave being injected into the string were some higher integer multiple of a half wavelength (referred to as a "h*armonic*" of the "*fundamental*" half wavelength) such as 3/2 or 5/2 or 10/2 wavelengths (etc.), then after it has traveled twice the distance of the string (i.e. out and back), it has gone an even multiple of a wavelength (e.g. 2 x 3/2 = 6/2 = 3 wavelengths), and is therefore back in phase with the next wavefront, reinforcing its energy once again. It is for this reason that musical instruments produce a fundamental tone, *plus* multiples of that fundamental (the higher "harmonics"). Each type of instrument tends to have its own characteristic response to these harmonics, giving each instrument its own unique sound (which explains why a "C" sharp[53] on a violin sounds uniquely different from that same "C" sharp on a trumpet, etc.).

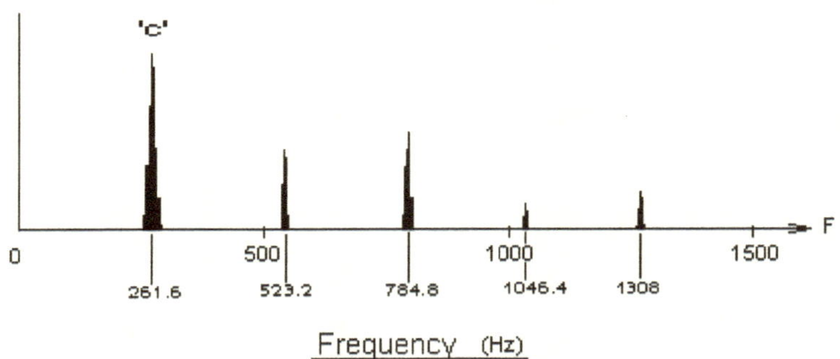

Figure A.2 An audio spectrum of a typical musical instrument, including a few harmonics. Notice the varying amplitudes of the successive harmonics, with the amplitude of each harmonic being affected by the instrument's unique response to each of these frequencies.

As discussed in the body of the text, individual atoms and molecules also develop well-defined resonant modes, each being entirely a function of energy stored and/or energy released by that type of atom. When an external object imparts the correct amount of energy into these objects to match a specific Quantum energy transition, it effectively strikes a "resonant chord" in the atom or molecule, placing it into a well-defined, discrete, resonant mode.

---

[53] Each musical note in our twelve note scale, has a specific frequency according to the following formula: $F(n) = 27.5 \times 2^{(n/12)}$. For example, a mid-octave "C" is 261.625 Hz (i.e. n=39), while a mid octave "F" is 349.228 Hz (n=44).

We point out that whether we are describing a macro object (such as a steel bar, or string) or a Quantum object (e.g. an atom, or molecule), the resonant excitation process is a function of energy being absorbed and stored in that object through this resonance effect.

As the waves travel through each object (e.g. steel bar, violin string, etc.), we note that the mass of the object itself does not migrate, but simply provides a medium through which the *energy* of the waves themselves migrate (very analogous to an ocean wave). If it were actually chunks of mass that migrated to create resonance, anytime two "pulses" of mass met at some point in the middle of the string (or in the atomic "energy-cloud-thing" that is the electron), they would collide and dissipate energy in the process. What propagates in each of these systems is energy itself, and therefore when two pulses of energy meet in the middle of a system, they pass right through one another completely "unscathed"/undiminished by the process.

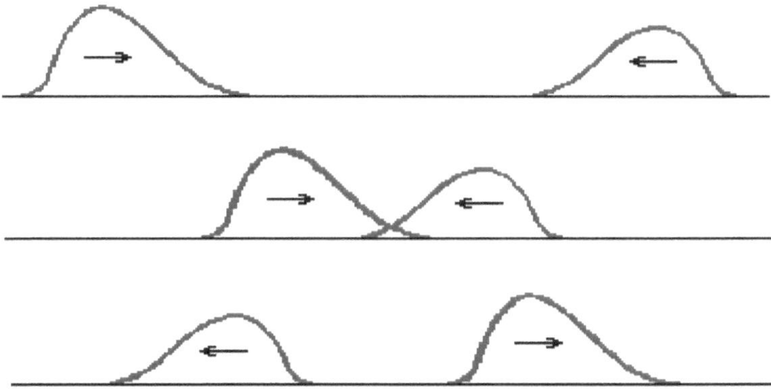

Figure A.3 Two pulses of energy propagating in opposite directions passing through each other unaltered.

The fundamental resonant frequency of a tight string depends on the speed at which energy propagates through that string, which in turn depends on the tension in the string (T) and the density (ρ, mass per unit length) of the string:

$$F_1 = \mathrm{SQRT}\,(T / \rho)$$

In other words, the tighter we pull the string, the faster the wave will propagate (and hence fitting "more wavelengths" for the same length of string, implying a higher resonant frequency). We also see that the thicker the string is (i.e. the greater its density), the lower will be its resonant frequency (since increasing the thickness of the string tends to slow the propagation of the wave down the string).

As mentioned, the length of the string coupled with the speed at which energy propagates through the string, determines the resonant wavelength of the string:

$$L = n \lambda / 2$$

For the fundamental frequency:

$$\lambda = 2 L \qquad \text{(fundamental, } n=1\text{)}$$

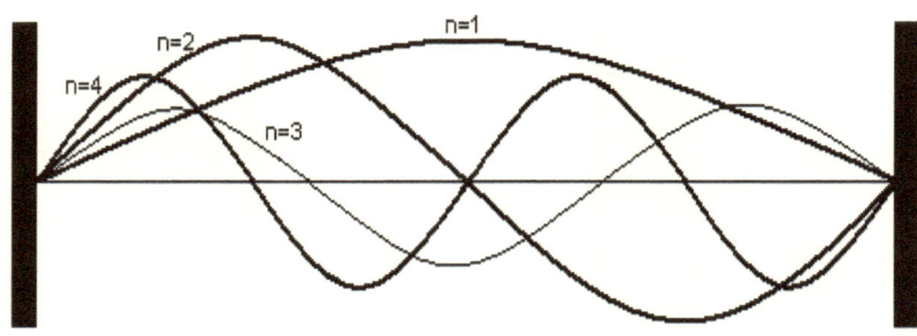

Figure A.4 Waves supported on a string

Since frequency and wavelength are related as:

$$\mathrm{Freq} = v / \lambda$$

Therefore:

$$\text{Freq} = \text{SQRT}(T / \rho) \;/\; 2L$$

Example: Violin string resonance

Given a violin string 0.3 m long, with a density of 1g/m under a tension of 25N, find its fundamental resonant frequency.

$$\text{Freq} = \text{SQRT}(T / \rho) \;/\; (2L)$$

$$= \text{SQRT}(25\text{N} / 0.001 \text{ kg/m}) \;/\; 2(0.3 \text{ m})$$

$$= 263.5 \text{ Hz}$$ (roughly a mid-octave "C")

Example: Fretting

When we "fret" a string against the neck of the instrument (e.g. a guitar or banjo), we effectively change the length of the string, and thereby change its resonant frequency. How long does this same string need to be when we "fret" it to generate a mid-range "E" (~330 Hz)?

$$\text{Freq} = 330 \text{ Hz} = \text{SQRT}(25\text{N} / 0.001 \text{ kg/m}) / (2L)$$

Therefore:

$$L = \text{SQRT}(25\text{N} / 0.001 \text{ kg/m}) \;/\; 2(330 \text{ Hz})$$

$$= \sim 0.24 \text{ meters}$$

A pipe closed at one end and open at the other is similar to the string, except that by closing one end and opening the other, we have two completely opposite "impedances" in our media at the two ends. On the closed end, the wall prevents the air from moving or vibrating immediately at the face of the wall, and therefore the wave must "null" or have zero vibration (the same as it did at the ends of the string). At the open end however, the wave encounters exactly the opposite "impedance" conditions (i.e. much greater freedom of movement of air), and as a result the wave at the opening of the pipe is at its strongest intensity (i.e. at a "peak" or "anti-node"). As a result, the wave developed inside the pipe is a quarter wave, rather than a half wave. Hence the Length of the pipe supports a quarter wave of its fundamental frequency:

$$L = n\lambda / 4 \quad \Rightarrow \quad 1 * \lambda / 4$$

(open pipe fundamental)

$$\text{i.e.} \quad \lambda = 4L$$

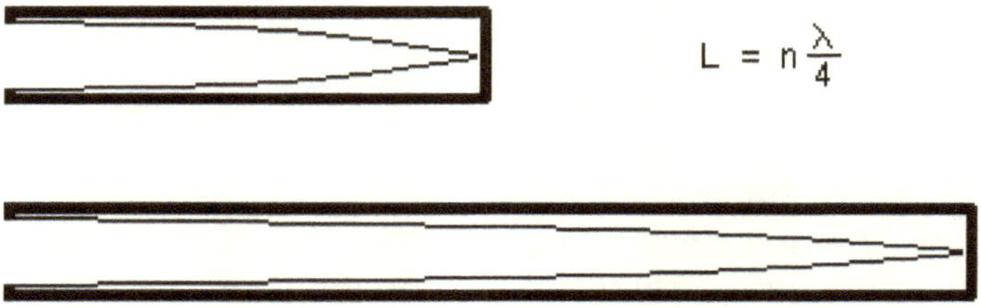

$$L = n\frac{\lambda}{4}$$

Figure A.5 Resonance within an open pipe.

Example: penny whistle

Given v = 345 m/s and an empty pipe 11.65 centimeters long with one end closed, if we blow across the open lip of the pipe, at what frequency will the pipe resonate?

$$\text{Freq} = v / 4L = 345 / (4 \times 0.1165 \text{ m})$$

$$= 740 \text{ Hz}$$

(middle "G" flat)

Appendix B: Fourier Mechanics

Back in the late 1800's, scientist began to realize that waves and resonance effects were everywhere around us, and indeed were an integral part of everything that exists in the universe (e.g. the sounds of our voices, the vibrations on our ear drums, the undulations of a tree in a stiff breeze, the tones created by a musical instrument, light traveling across the universe, etc.). In order to understand these waves, we needed a set of mathematical tools that were particularly adapted for use in modeling and analyzing waves.

To address this need, a rather clever guy named Jean Baptist Fourier developed something of an "algebra" of waves, which became known as "Fourier Mechanics". A number of excellent books have been written on the subjects of Fourier Superpositions, Convolutions and Transforms, as well as the "computer friendly" transform known as the "Fast Fourier Transform". We will recap some of the more significant highlights of Fourier Mechanics here, but those who find these techniques intriguing, would be well rewarded by reviewing our brief tutorial of the Fast Fourier Transform on our website "EpiphanyBySteveLee.com" (misc. tab), or by spending some time pursuing these topics more thoroughly at their local university library (e.g. Arfken [3], etc.).

In essence, Fourier Mechanics is the language of waves. Through it we can dissect or analyze anything that has (or can be portrayed to have) a wave-like nature. Propagating E&M waves are ideal candidates for these techniques, but they can be equally exploited to analyze an electronic system's characteristic response, the components of a waveform, or the molecules that our E&M waves propagate through, etc.

To begin our overview of Fourier Mechanics, let us revisit the subject of Superposition.

Fourier Superposition:

The basic premise of Fourier Superposition is that we can model any "well-behaved" periodic signal or structure using a well-chosen set of Orthogonal functions as our basic building blocks (i.e. as a "basis set"). These can be a collection of Sine wave harmonics (a fundamental frequency and multiples of that fundamental), or a similar set of Cosines. Other possibilities include Bessel functions, Spherical Harmonics, Legendre Polynomials, etc. (Arfken [3] is a suitable book to cover most of these functions).

At first glance, these functions may seem a bit intimidating, but no more so than the Sine() or Cosine() functions. We do amazing things with them, so much so that they have transcended the level of merely being useful tools, to the point where they now form the indispensable language and structure of many scientific disciplines, including most modern radio communications techniques (e.g. CDMA and LTE), as well as Quantum Mechanics, which provides us the next level of understanding about the universe beyond Maxwell's Equations (see Chapter 6). Max Plank used Superposition to dissect the conundrum of Black Body radiation (ca. ~1901), and in the process struck a tiny crack in the study of science that soon opened up into the giant chasm of Quantum Physics that defines the underpinnings of everything in the known universe (not to mention the more "pedestrian" topics of chemistry, lasers, semiconductors, etc.).

Since most technicians and Engineers have some affinity to Sine and Cosine waves, we will concentrate on their use as our basis set for the following examples. Since orthogonality is a prerequisite of the Fourier basis sets, we note the orthogonality of Sine and Cosine waves can be expressed as:

$$\int_{-T/2}^{+T/2} \mathrm{Sin}(m\,\omega\,t)\,\mathrm{Sin}(n\,\omega\,t)\,dt = T/2 * \delta_{(m,n)}$$

$$\int_{-T/2}^{+T/2} \mathrm{Cos}(m\,\omega\,t)\,\mathrm{Cos}(n\,\omega\,t)\,dt = T/2 * \delta_{(m,n)}$$

where "$\delta_{(m,n)}$" is the "Kronecker delta" defined in Chapter 3 (see Figure 3.4 for a graphical representation of the Sine() orthogonality relation above). With the orthogonality requirement met for our basis functions, we can use our Sine waves to construct virtually any [well behaved] function we want.

Example B.1 In Chapter 3, we demonstrated the construction of a square wave using a collection of "N" Sine waves:

$$s(t) = \sum_{n=0}^{N} a_n \, \text{Sin}(n \, \omega \, t) \qquad \text{(with } \omega = 2 \pi f\text{)}$$

Which we will now justify using the following form for our Fourier Series representation of s(t):

$$s(t) = a_o/2 + \sum_{n=1}^{\infty} a_n \, \text{Cos}(n \, \omega \, t) + \sum_{n=1}^{\infty} b_n \, \text{Sin}(n \, \omega \, t)$$

$$\text{with} \quad a_n = 2/T \int_{-T/2}^{+T/2} s(t) \, \text{Cos}(n \, \omega \, t) \, dt$$

$$\text{and} \quad b_n = 2/T \int_{-T/2}^{+T/2} s(t) \, \text{Sin}(n \, \omega \, t) \, dt$$

(where T = period of our waveform). To prove the statement for the coefficients, multiply both sides of the Fourier Series representation by Cos(m ω t) for the a_n's and multiply both sides by Sin(m ω t) for b_n's, and integrate over the period "T". We will show the proof for the a_n's:

$$\int_{-T/2}^{+T/2} s(t) \, \text{Cos}(m \, \omega \, t) \, dt = \int_{-T/2}^{+T/2} a_o/2 \, \text{Cos}(m \, \omega \, t)$$

$$+ \int_{-T/2}^{+T/2} \sum_{n=1}^{\infty} a_n \, \text{Cos}(n \, \omega \, t) \, \text{Cos}(m \, \omega \, t)$$

$$+ \int_{-T/2}^{+T/2} \sum_{n=1}^{\infty} b_n \, \text{Sin}(n \, \omega \, t) \, \text{Cos}(m \, \omega \, t)$$

If we evaluate the $a_o/2$ integral, we find it reduces to zero (where one period "T" = 2π). Furthermore, since Sine and Cosine waves are orthogonal over a full period, the b_n integral also goes to zero, leaving only the a_n terms. We further note that Cos(n ω t) Cos(m ω t) = (T/2) * $\delta_{(m,n)}$ (i.e. the only time the integral is *not* zero is when n=m), hence the remaining integral on the right side reduces to: a_n T/2, leaving us with:

$$\int_{-T/2}^{+T/2} s(t)\,\mathrm{Cos}(m\,\omega\,t)\,dt \quad = \quad a_n\,T/2$$

Therefore:

$$a_n \;=\; 2/T \int_{-T/2}^{+T/2} s(t)\;\mathrm{Cos}(n\,\omega\,t)\;dt$$

(with a similar proof for our b_n coefficients). Armed with these equations for our coefficients, we can now construct our Fourier Series representation for a square wave (asymmetric about 0, with period "T" and amplitude range of +1 and -1). We start with the "a_n" coefficients, and split the integral into two parts around zero:

$$a_n \;=\; 2/T\{ \int_{-T/2}^{0} s(t)\,\mathrm{Cos}(n\,\omega\,t)\;dt + \int_{0}^{+T/2} s(t)\;\mathrm{Cos}(n\,\omega\,t)\;dt \}$$

since s(t) is asymmetric and Cos() is symmetric, both integrals reduce to zero (which tells us that in an asymmetric waveform, there are no "a_n" coefficients in the Fourier series, since with only Cosine terms in "a_n", the "a_n" coefficients represent the "symmetrical components" in the series). This can easily be verified by performing the integrations while noting that we defined our waveform amplitudes such that s(t) = -1 preceding zero, and s(t) = +1 afterwards.

Similarly, we find that the "b_n" coefficients represent the "asymmetrical components" of our waveform, and since our waveform s(t) is by definition asymmetric about zero, we do not expect these coefficients to be zero:

$$b_n \;=\; 2/T\;\{ \int_{-T/2}^{0} s(t)\;\mathrm{Sin}(n\,\omega\,t)\;dt \;+\; \int_{0}^{+T/2} s(t)\;\mathrm{Sin}(n\,\omega\,t)\;dt \}$$

$$=\; 2/T\;[\int_{-T/2}^{0} (-1)\;\mathrm{Sin}(n\,\omega\,t)\;dt \;+\; \int_{0}^{+T/2} (+1)\;\mathrm{Sin}(n\,\omega\,t)\;dt\,]$$

$$=\; 2/T\;\;(1/n\omega)\;(\mathrm{Cos}(n\,\omega\,t)\;|^{0}_{-T/2} \;-\; 2/T\;\;(1/n\omega)\;(\mathrm{Cos}(n\,\omega\,t)\;|^{T/2}_{0}$$

$$=\; 1/(n\,\pi)\;\{\,1 - \mathrm{Cos}(-n\,\pi) \;-\; [\,\mathrm{Cos}(n\,\pi) \;-\; 1\,]\,\}$$

(where we used the fact that $\omega = 2\pi / T$). Therefore, our "b_n" coefficients reduce to:

$$= 1/(n\pi)\ \{ 2 - 2\text{Cos}(n\pi) \}$$

$$= 2/(n\pi)\ \{ 1 - \text{Cos}(n\pi) \}$$

Therefore, we find that:

$b_n = 4/(n\pi)$ if "n" is odd, and:
$b_n = 0$ when "n" is even.

Therefore, our Fourier Series representation for our asymmetric square wave is:

$$s(t) = (4/\pi)\ \text{Sin}(\omega t) + (4/3\pi)\ \text{Sin}(3\omega t) + (4/5\pi)\ \text{Sin}(5\omega t)\ \ldots$$

Using the same general concepts embedded in the Fourier approach, we could similarly use it to model any number of phenomenon, including the currents in a circuit, or the radiation pattern generated off an antenna, etc. In this latter example, we could use a collection of basic radiators (such as dipoles) as our basis "building blocks", each with a different current phase and/or amplitudes. Similarly, we could use this approach to construct a model of an atom or molecule using similar orthogonal basis functions "$\Phi_{k\,(r)}$":

$$\Psi(r,t) = \sum a_k(t)\ \Phi_{k\,(r)}$$

etc.

Fourier Transform:

Since there is a relationship between time and frequency (e.g. F = 1/t in sine and cosine waves), we find we can learn a great deal more about a system under study if we look not only at its behavior over time, but also at the frequency components of that system exhibited over that same period of time. To switch between the temporal perspective and its related spectral perspective, Fourier developed a mathematical tool in his "tool box" known as the "Fourier Transform". Engineers who are familiar with the Laplace transform know that such a transform technique can vastly simplify our analysis. (Those who have used the Laplace transform will note that the Fourier transform is closely related to the Laplace transform, with the "slight" modification of an extra "i").

The transform between these two views is possible since both are, after all, describing exactly the same distribution of energy in a given system, only from two different perspectives (time "t", and frequency "ω" = 2 π f):

$$S(\omega) = \int_{-\infty}^{\infty} s(t) \exp(-i\,\omega\, t)\, dt$$

$$s(t) = \int_{-\infty}^{\infty} S(\omega) \exp(+i\,\omega\, t)\, d\omega$$

In other words, the energy in a system is not simply a "signal" that varies with time, but rather a distribution of energy which typically exhibits something of a spectral "fingerprint" that reflects the resonance characteristics of that system. Therefore, whether we describe this distribution as a flow of energy verses time, or as a distribution of energy across frequency, we are describing the same thing. Mathematically, this "duality" is codified by what is known as the "Bessel-Parseval" relation:

$$\int_{-\infty}^{\infty} s(t) \exp(-i\,\omega\, t)\, dt = \int_{-\infty}^{\infty} S(\omega) \exp(+i\,\omega\, t)\, d\omega$$

where S(ω) is the signal as a distribution in frequency, and s(t) is the signal as a function of time. The "Bessel-Parseval" relation tells us that the information contained in our energy distribution is the same whether we are looking at it as an "oscilloscope trace" over time, or as "spectral plot" of its various frequency components. The Bessel-Parseval relation should therefore seem fairly reasonable, since we would expect the energy in our system to be the same, regardless of which perspective we happen to choose to view.

As examples of the Fourier Transform technique, we will now show the transformation of several basic functions, beginning with a simple delta function. We note that the delta function is particularly easy to transform, since it is zero everywhere except at one location. This effectively collapses the integral into a single value at that point:

Example B.2 five basic functions and their corresponding Fourier transforms (see Figure B.1).

- XFM[$\delta_{(t,0)}$]:

$$\int_{-\infty}^{\infty} \delta_{(t,0)} \exp(-i\ \omega\ t)\ dt \quad = \quad \exp(0) \quad = \quad 1$$

- XFM[1 bit, amplitude = "A"]:

$$\int_{-T/2}^{T/2} A\ \exp(-i\ \omega\ t)\ dt \qquad \text{(using eqn. D.1):}$$

$$= \int_{-T/2}^{T/2} A\ \mathrm{Cos}(\omega\ t)\ dt \ - \ i \int_{-T/2}^{T/2} A\ \mathrm{Sin}(\omega\ t)\ dt$$

$$= A\ \mathrm{Sin}(\omega\ t)\ \Big|_{-T/2}^{T/2} \ - \ 0 \quad \text{(due to symmetry)}$$

$$= 2A\ \mathrm{Sin}(\omega\ T/2)\ /\ \omega \quad = \quad \text{"Sinc" function}$$

- XFM[$\exp(-i\ \omega_a\ t)$]:

$$\int_{-\infty}^{\infty} \exp(-i\ [\omega_a + \omega]\ t)\ dt \quad = \quad \mathrm{Delta}\ (\omega_a + \omega)$$

- XFM[$A\ \mathrm{Cos}(\omega_a t)$]:

$$\int_{-\infty}^{\infty} A/2\ [\ \exp(i\ \omega_a\ t)\ +\ \exp(-i\ \omega_a\ t)\ ]\ \exp(-i\ \omega\ t)\ dt$$

$$= A/2 \int_{-\infty}^{\infty} [\ \exp(-i\ [\omega - \omega_a]\ t) + \exp(\ -i\ [\omega + \omega_a]\ t)\ ]\ dt$$

$$= A/2\ \mathrm{Delta}\ (\omega - \omega_a\ )\ +\ A/2\ \ \mathrm{Delta}\ (\omega + \omega_a\ )$$

where we used Euler's relation (eqn. D.1) from Appendix D to rewrite our Cosine term:

$$\mathrm{Cos}(xt) \ = \ ½\ (\ \exp(ixt)\ +\ \exp(-ixt)\ )$$

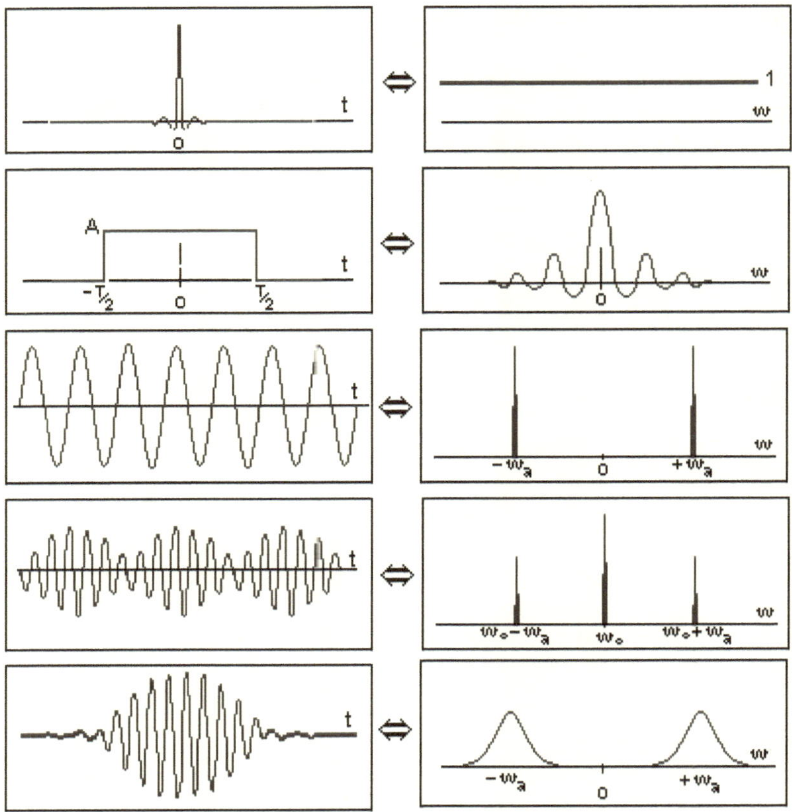

Figure B.1 Five functions and their Fourier transforms.

From Euler's relation, the orthogonality between Sines and Cosines (see equation 3.1), and the orthogonality between the harmonics of both Sines and Cosines (see equation 3.2), we can state that:

$$\int_{-\infty}^{\infty} \exp(i\,n\,x)\ \exp(i\,m\,x)\ dx \;=\; \delta_{(n,m)}$$

- XFM[$\exp(-\omega^2)\,\mathrm{Cos}(\omega_a t)$]:

$$= \tfrac{1}{2}\ {}_{-\infty}\!\int^{\infty} \exp(-\omega^2)\ [\ \exp(i\ \omega_a\ t) + \exp(-i\ \omega_a\ t)\ ]\ \exp(-i\ \omega\ t)\ dt$$

$$= \tfrac{1}{2}\ \exp(-\omega^2)\ \ {}_{-\infty}\!\int^{\infty}\ [\ \exp(-i\ [\omega - \omega_a]\ t) + \exp(\ -i\ [\omega + \omega_a]\ t)\ dt$$

$$= \tfrac{1}{2}\ \exp(-\omega^2)\ \{\ \mathrm{Delta}\ (\omega - \omega_a\ )\ +\ \mathrm{Delta}\ (\omega + \omega_a\ )\ \}$$

$$= \tfrac{1}{2}\ \exp(-\omega_a{}^2)\ +\ \tfrac{1}{2}\ \exp(+\omega_a{}^2)$$

There are several very interesting results from these examples worth noting (see Figure B.1): In the first transform, we see that a perfectly narrow pulse has an infinitely broad spectrum, suggesting that the more we attempt to confine our pulse in time, the more spectral components it will have (note that a similar effect occurs spatially, as discussed in Chapter 5; see Figure 5.15).

The second example shows that the transform of a single bit of amplitude "A" and width "T" has a spectral signature which is rich in harmonics well above the main body of the fundamental's energy, further supporting our conclusions derived from our previous Fourier Superposition example of a square wave (see Figures 3.5 and 3.6). Here too we can see that the abrupt edges in the square bit corresponds to a rich spectrum of harmonics.

Finally, in the last transform in the previous example, we see that a Gaussian distribution in time-space, remains a Gaussian distribution after transforming into spectral space. This highlights a rather unique characteristic of Gaussian distributions, in that they are one of the few types of distributions which retain their form when transformed between time and frequency (i.e. their distribution in time is characteristically similar to their distribution of frequencies). As a consequence, pulses which are composed of a Gaussian distribution of frequency components experience very little "smearing" as they propagate through a system over time (we will revisit this result again in the discussion which follows).

Fourier Mechanics and Modulation:

The Fourier Transform can be applied to a number of different wave-related processes, including the concept of modulating an RF or laser carrier. The application of Fourier tools in this case yields a great deal of useful insight about modulation in general.

If we define the Fourier transform pairs of:

XFM [s(t)] <=> S(ω) and
XFM[m(t)] <=> M(ω)

being the transform relations between the time domain and the frequency domain, we can describe the process of modulation in a very eloquent way using nothing more than the language of a Fourier Mechanics.

To do this we once again make use of the Euler relation (see Appendix D) to rewrite the Cosine function in terms of exponentials: cos(xt) = ½ (exp[ixt] + exp[-ixt]). This will allow us to show that the modulation of a signal s(t) onto a carrier at frequency "ω_c" produces a "sideband" on either side of the carrier:

We prove this using Fourier Transforms as follows:

$$\text{XFM}\,[s(t)\,\text{Cos}(\omega_c\, t)\,]$$

$$= \text{XFM}\,[\tfrac{1}{2}\; s(t)\exp(i\,\omega_c\, t)\; + \;\tfrac{1}{2}\, s(t)\exp(-i\,\omega_c\, t)\,)\,]$$

$$= \text{XFM}\,[\tfrac{1}{2}\; s(t)\exp(i\,\omega_c\, t)\,] + \text{XFM}\,[\;\tfrac{1}{2}\, s(t)\exp(-i\,\omega_c\, t)\,)\,]$$

$$= \tfrac{1}{2}\;{}_{-\infty}\!\int^{\infty} s(t)\exp(i\,\omega_c\, t)\;\; \exp(-i\,\omega\, t)\, dt$$

$$+ \;\tfrac{1}{2}\;{}_{-\infty}\!\int^{\infty} s(t)\exp(-i\,\omega_c\, t)\;\; \exp(-i\,\omega\, t)\, dt$$

Therefore:

$$s(t)\,\text{Cos}(\omega_c\, t) \;=\; \tfrac{1}{2}\, S(\omega - \omega_c) \;+\; \tfrac{1}{2}\, S(\omega + \omega_c)$$

This shows one central peak at our carrier frequency (ω_c), and two "sideband" peaks of ½ amplitude on either side of the carrier, with displacement being proportional to the frequency of the input signal (ω) (see Figure B.2). Note that as the frequency of the input signal increases, the sidebands move farther away from the carrier. It is for this reason that the signal(s) being fed into the modulator typically are first put through a Low Pass filter, to remove any higher frequency components that would otherwise create out-of-channel spurious emissions.

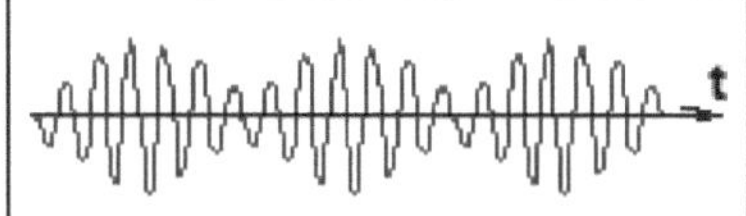

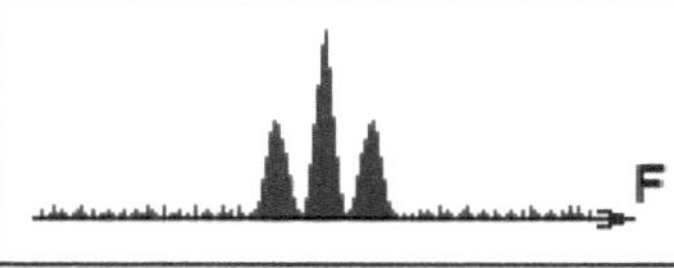

Figure B.2 Modulated carrier with two sidebands.

To demodulate an RF carrier at the receiving end, we simply multiply the received signal by a locally generated "monochromatic" Cosine wave (i.e. the output of our Local Oscillator, "LO"). We represent our modulated signal as: $s(t) = m(t)\,\mathrm{Cos}(\omega_{RF}\,t)$, and then multiply this by the LO Cosine wave:

$$s(t) * \mathrm{Cos}(\omega_{LO}\,t) = m(t)\,\mathrm{Cos}(\omega_{RF}\,t) * \mathrm{Cos}(\omega_{LO}\,t)$$

If again we substitute the Euler identity for each of the Cos() terms and multiply them out, this becomes:

$$= \tfrac{1}{2}\, m(t)\,[\exp(i\,\omega_{RF}\,t) + \exp(-i\,\omega_{RF}\,t)] * \tfrac{1}{2}\,[\exp(i\,\omega_{LO}\,t) + \exp(-i\,\omega_{LO}\,t)]$$

$$= \tfrac{1}{4}\, m(t)\,\{\exp(i\,[\omega_{RF} + \omega_{LO}]\,t) + \exp(i\,[\omega_{LO} - \omega_{RF}]\,t) + \exp(i\,[\omega_{RF} - \omega_{LO}]\,t) + \exp(i\,[-\omega_{LO} - \omega_{RF}]\,t)\}$$

Using the definition for our transform:

$$\mathrm{XFM}[X(\omega)] = \int [x(t)\exp(-i\,\omega\,t)]\,dt$$

we get:

$$\text{XFM}\,[s(t) * \text{Cos}(\omega_{LO}\, t)]$$
$$= \tfrac{1}{4} \int [\, m(t) \exp(i\, [\omega_{RF} + \omega_{LO}]\, t)\ \exp(-i\, \omega\, t)\,]\, dt\ +\ \ldots$$

$$= \tfrac{1}{4} \{\ M(\omega_{RF} + \omega_{LO}) + M(\omega_{RF} - \omega_{LO}) + M(-\omega_{RF} + \omega_{LO}) + M(-\omega_{RF} - \omega_{LO})\ \}$$

If we choose $\omega_{LO} < \omega_{RF}$, the last two terms fall into the "negative" frequency domain, and thus have no corresponding physical interpretation. Therefore, out of our modulator we have the two original input signals (the RF input signal and the LO Cosine wave), plus the two remaining product terms shown above: the *sum* and the *difference* terms:

$$RF + LO + \tfrac{1}{4} M(\omega_{rf} + \omega_{LO}) + \tfrac{1}{4} M(\omega_{rf} - \omega_{LO})$$

The goal of the mixing process is to "down-convert" our modulated signal from the higher RF tuner range to a lower frequency (i.e. our "Intermediate Frequency", or "IF"), making the processing electronics designed around the IF easier to build and operate.

To eliminate everything in the above equation except the difference term (i.e. the "down-converted", IF signal), all we have to do is place a low pass filter immediately after the Mixer stage. We then pass this IF energy through an amplifier tuned specifically for the IF and demodulate the result to strip out the original information sent by the transmitter.

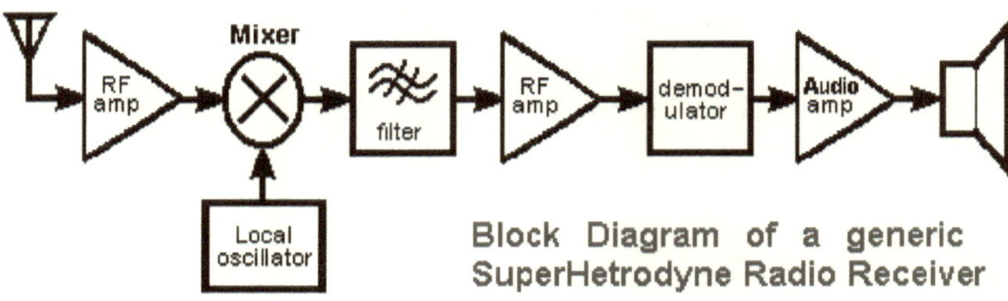

Block Diagram of a generic SuperHetrodyne Radio Receiver

Question: What would happen if we used 10.0 MHz (WWV broadcast channel in Fort Collins, Colorado) for our IF?

(When designing this type of receiver, the Local Oscillator frequency is typically chosen such that the IF is not near the operating frequency of any commonly used high power RF channels – e.g. all commercial broadcast frequencies, Short Wave, Emergency services, and two-way bands, with 10.7 MHz being a common choice for VHF receivers. Note also that receivers designed to operate at "ultra" high frequencies may use several such "down-conversion" stages, and therefore have several Local Oscillator frequencies and several IF stages, each chosen to avoid any commonly used radio link frequencies, such as FM stations, cellular base stations, FAA bands, etc.)

Fourier Convolutions:

The Fourier Convolution is the integral of the product of two distributions "f_a" and "f_b":

$$F^{ab}_{(t)} = \int_{-\infty}^{\infty} f_{a(\omega)} \, f_{b(t-\omega)} \, d\omega$$

A convolution can therefore be thought of as multiplying narrow slices of each distribution function together as they pass through each other over the full range of integration and summing the results.

The resulting output of the convolution effectively "inherits" the characteristics of the two inputs distributions. This can be depicted using a sequence of "cartoon" snapshots, passing one function through the other to generate the product in the process (see Figure B.3). If either function goes to zero at any point along the range, the product is zero (we therefore only need to consider the limited range where both are non-zero, rather than over the full range of $\pm\infty$). Examples of two functions convolved together include a test pulse distribution passed through a function representing a system band-pass, or a light pulse in a fiber optic cable passing through a reflection function that represents a defect in the cable, (etc.).

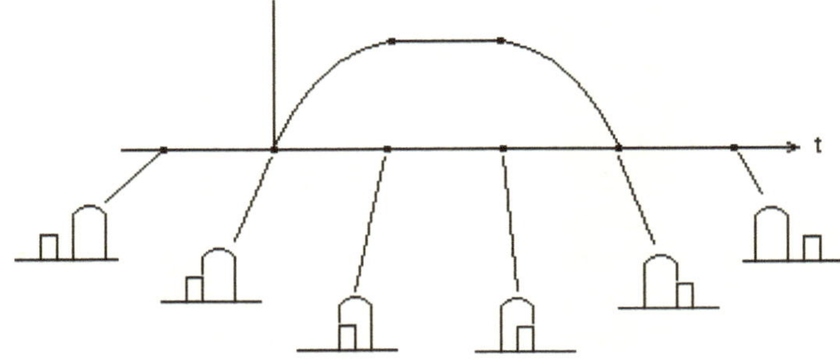

Figure B.3 Example convolution, in "cartoon" snapshot form (moving left-to-right in the figure), with the product taking on the shape/width characteristics of the two input functions, depicted here as the large "hump" above the "t" axis.

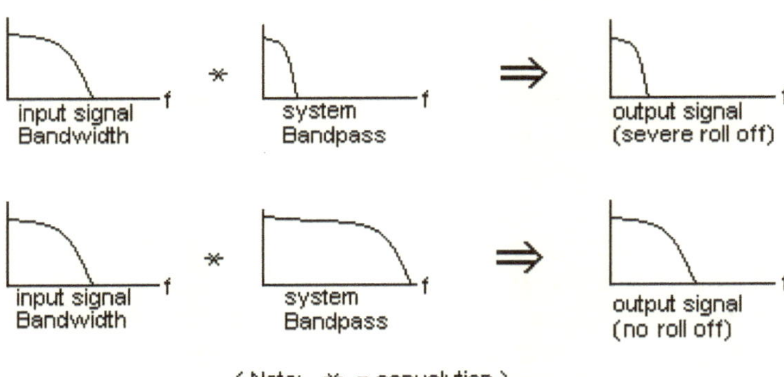

Figure B.4 Sample convolution: Bandwidth vs Spectral Content.

Dispersion:

Anytime a signal containing multiple frequency components is processed through a system that acts on each frequency component *differently* (e.g. different wavelength light through a section of glass, or fiber, etc.), each frequency component of that signal experiences a different gain and/or delay as it moves through that system. This frequency dependent response is known as "Dispersion".

In the case of photons passing through an optical system, dispersion is ultimately due to the response of the atoms and/or molecules in the system, which as we saw in Chapters 2 and 3, tends to be highly wavelength dependent. What's more, we find as a consequence of the "Uncertainty Principle" ($\Delta E \, \Delta t \geq 2h\pi$), that it is impossible to create a purely monochromatic source (e.g. RF oscillator, laser, etc.) with a single wavelength of light. Though the Quantum aspects of atoms indicate that the photons released by an excited atom may be very wavelength specific ($E = h f$), there are a number of other effects and mechanism that affect the frequency of energy being released by the source (e.g. Doppler shifting, collisional broadening, thermal variations, etc.), that tend to spread the wavelength range of the energy emitted by any source.

As a result, no oscillator or light source (including the most refined laser available) will produce a perfectly narrow, single wavelength emission. In fact, reality (entropy) being what it is, we find that most sources used in everyday applications tend to be very far from the ideal, often producing emissions with not only a broad spectral peak, but often with more than one such spectral peak, which in a long-haul fiber optic link can introduce considerable dispersion related ISI problems (e.g. see Figure 8.5). As these various frequency components travel outward from the source, they each travel with a slightly different velocity through the propagating medium, be that in the atmosphere, a block of plastic, a long glass fiber, etc. Consequently the "faster" components in the pulse tend to lead the "slower" components, spreading out the pulse as it travels.

If we define the "Group Velocity" of our wave packet "Vg" to be the overall "average" speed at which this packet of frequencies (energies) propagates out into space[54], and assume

---

[54] This definition of group velocity as the overall speed at which a wave packet's energy travels is true in all cases where the dispersion is not extreme. However in the extreme case, this generalization is no longer valid, since as the separation of the various frequency components becomes so severe, the wave packet or "group" ceases to

the wave velocities within the group are all proportional to frequency, then the spread of velocities is proportional to the spread of k (where $k = 2\pi/\lambda$):

$$\Delta(Vg) \approx \Delta k \qquad \text{or}$$

$$\Delta(Vg) \approx 2\pi / \Delta\lambda \qquad \text{or}$$

$$\Delta(Vg) * \Delta\lambda \approx 2\pi$$

The range of values in velocity and wavelength represented here as $\Delta(Vg)$ and $\Delta\lambda$ represent the uncertainty in these variable, and reflect the fact that there is some ambiguity as to the wavelength and speed of the wave packet as it travels. Since a wave packet is not like a car or a bullet with well-defined boundaries, but rather more of a bundle of frequencies, the movement of these packets are inherently nebulous by nature. By analogy a sound or ocean wave is similar, in that they have energy which extends some distance ahead of and behind the main body of the wave as a leading and trailing "swell". E&M pulses are even less well-defined, with their higher frequency components traveling slightly faster than the lower frequency components. As a result, the higher frequency components ("Brillouin Precursors") tend to lead the main body of the pulse, arriving slightly ahead of the lower frequency components [4].

The alert reader may have noticed at this point that the above relationship between uncertainties in our propagating pulse is remarkably similar to the "Heisenberg Uncertainty Principle" discussed in Chapter 3. The similarity is due to the fact that this E&M principle is a natural outgrowth of the wave nature of these electromagnetic pulses, as the Heisenberg Uncertainty Principle is an outgrowth of the wave nature of electrons and atoms. The ambiguity inherent in this wave-nature of Quantum objects in fact is the source of much of the seemingly bizarre or "counter-intuitive" aspects of Quantum Physics equations and theorems. On the plus side however, it is

---

remain intact as a single entity, and as such the transport of energy "fragments" into a function of multiple separate "sub-groups".

this wave-like nature that allows us to bring to bear all the Fourier tools (e.g. superposition of wave-like basis functions, as well as the Fourier Transform to toggle between various state function "spaces", etc.), harnessing the extreme usefulness of these Fourier tools in the process.

In Chapter 3 we noted that the wave-like nature of Quantum objects created an inherent uncertainty as to certain measurable quantities (e.g. position, momentum, energy, wavelength, etc.) of atoms, molecules, electrons, etc. In Quantum literature this is typically discussed under the "Heisenberg Uncertainty Principle", however we noted that even classical wave-like objects have this same inherent uncertainty to them. In other words, the uncertainty in group velocity of all wave-like objects:

$$\Delta(Vg) * \Delta\lambda \geq 2\pi$$

is just another reflection of this same characteristic nature of waves.

We note that though this equation states the product of uncertainties is "greater than or equal to" 2π, the only time the above uncertainty product actually approaches 2π is when the distribution of waves in the wave packet is a Gaussian distribution. In that case, the wave packet propagates with little or no dispersion, as described mathematically in the last Fourier Transform shown in example B.2 (see Figure B.1). We should mention however that this "ideal" Gaussian case is generally not what we find in the typical "square" bit pulses used in driving the average fiber optic laser. In that case, the laser pulses sent down a fiber link are typically composed of abruptly transitioned pulses very unlike the Gaussian type (i.e. very abrupt leading and trailing edges of each pulse, typically containing significant high frequency components). As a result, these typical cases tend to suffer a fair amount of "chromatic" dispersion as they travel, with the product of wavelength and velocity uncertainties often being considerably higher than the minimum 2π shown in the above equation.

Exercise B.1 By examining the mathematical definition of the Fourier Transform pair, it is obvious that they form a symmetrical set. For example, we found in example B.1 that when we transform a delta pulse in time, we get an infinitely wide spectrum in frequency space (ω). If we were to also transform a delta spike in frequency (i.e. a perfectly pure, single frequency spectral peak) we get a constant in time-space (i.e. the tone generator that produced that perfectly pure single frequency would have to have been running and stabilized for an infinitely long time). Use the same approach shown in example B.2 to prove this transform symmetry for all functions shown in example B.2. (i.e. transform the results in example B.2 in the opposite direction).

Exercise B.2 Find the Fourier Transform of the following functions: Sin(aωt), and a sequence of narrow bit pulses multiplied by Sin(ωt). What do you expect this last product to produce (hint: sampling an analog signal can be represented as a sequence of "wide" delta functions, since the sampling process is a periodic "snapshot" of the analog signal).

Exercise B.3 Find the inverse Fourier Transform of the following functions: Constant "A", Delta(ω-0) and a highly constrained (band-limited / filtered) band-pass spectrum "A" high and ω_a wide (similar to the 1 bit pulse in Example B.2, only in frequency space). What similarities are there to the results of Example B.2? What does this suggest?

Appendix C A mathematical model of a fiber optic cable:

Following the approach used by Kapron et al [15], we can model energy reflected back out of the fiber from a pulse of energy injected into it: If we define "R" as the characteristic Reflectance of the fiber (in dB), i.e.:

$$R_{(dB)} = 10 \log (P_{reflected}/P_{incident})$$

and "$\delta_{(t)}$" as the temporal width of a pulse launched down the fiber with amplitude "A", "v" as the group velocity of the pulse, and "dx" as the width of a small slice of fiber, then the reflection of the energy pulse off such a slice of fiber is:

$$dr_{(x)} = \delta_{(t-2x/v)} \, R \, A^2 \, dx$$

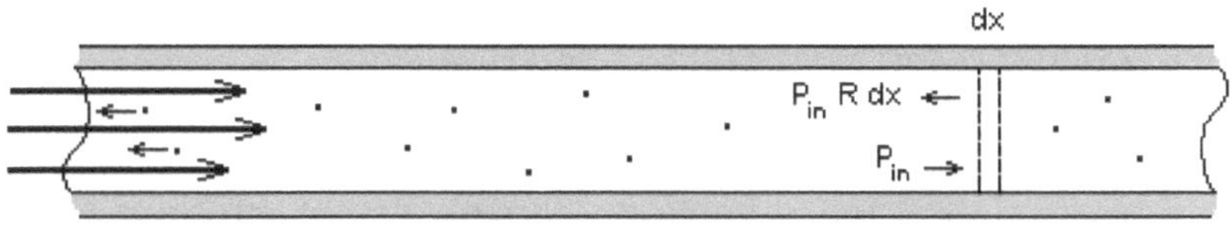

Figure C.1 Optic fiber reflectivity model.

Therefore, the total reflected energy is the integral of all reflections off all the tiny "slices" (dx) down the fiber (i.e. the sum of all energy from backscatter and/or reflection events):

$$r_{(t)} = \int_0^\infty \delta_{(t-2x/v)} \, R \, A^2 \, dx \equiv \text{"}G_{(t)}\text{"}$$

For compactness, we defined our integral term simply as a function "$G_{(t)}$". In essence, the energy reflected throughout the fiber is simply a function of the amount of energy injected into the fiber, coupled with the characteristic reflecting properties of the fiber itself. To simplify some of the following expressions, we will approximate our "reflectance" function "$G_{(t)}$" to:

$$"G_{(t)}" = R A^2 = R \exp(-\alpha v t)$$

where "α" is the characteristic attenuation coefficient for the fiber under test. We know that the more energy we inject into the fiber, the more energy is reflected. To increase the energy injected, we can either increase the amplitude ("A"), or we can increase the duration of the pulse ("D"). Therefore, considering the duration of the pulse, we see that the power ("P") reflected back out of the fiber is:

$$P_{out} = {}_0\int^D P_{in} G_{(t-\tau)} \, d\tau$$

$$= P_{in} R \; {}_0\int^D \exp(-\alpha v[t-\tau]) \; d\tau$$

$$= P_{in} R \; \exp(-\alpha v t) \; {}_0\int^D \exp(-\alpha v \tau) \; d\tau$$

(where "τ" is the pulse "on" time, and "t" is the pulse "time of flight" while moving through the fiber.)

If we define "r(x)" = reflection response, "P" = arbitrary pulse, "η" = backscatter coefficient, and "L" as the total length of the fiber link, then the reflections coming out of the fiber can be modeled by:

$$r(x) = \exp(-2\alpha L) \, \eta \, \exp(-\alpha v t) \; {}_0\int^L \exp(\alpha v \tau) \, P(t) \; R \, p(\tau) \; d\tau$$

Appendix D: Complex Numbers:

In the grand scheme of mathematical paraphernalia, there are several useful concepts which have been invented over the millennia that have proven to be seminal events in our quest for knowledge and understanding. Probably the most well known and least appreciated is the concept of zero, introduced to Western "mathematicians" (who previously thought it unnecessary) through various Arabic texts. There is some hint that this and other such mathematical entities may have been invented by some of the more ancient cultures such as the Egyptians, but the degradation by time and the elements make it difficult to peer too deeply back into the fog of time to be certain of its true origins.

We do know however that at least the ancient Egyptians, Indians, Chinese and possibly pre-Mayan cultures in the Americas were aware of a several such useful mathematical and Engineering "oddities" such as pi. In fact, the scant vestigial remains that survive from the Egyptian dynastic period in applied Mathematics, Engineering and Chemistry, still linger faintly in our collective cultures today – not the least of which is the possible reference to "π" in some ancient texts, as well as the embedded presence of "π" within the dimensions of the pyramids themselves. Other basic vestigial elements also persist around us today, including our word for Alchemy and Chemistry, derived from ancient Greek "Khemia" = "Egypt". A little off the subject, but interesting nonetheless.

The next mathematical "discovery" we would like to mention occurred in the 16th century, with the "invention" (or perhaps re-"invention") of negative numbers, which revolutionized mathematics yet again. Up until that time, the concept of a negative pig, or cow, or boat was considered tantamount to "demon speak" or just absolute fictitious non-sense. Slowly however, negative numbers gained acceptance as they proved extremely useful in solving certain quadratic and cubic equations, as well as being rather useful in such ordinary tasks as maintaining ledgers and balance sheets – not to mention keeping track of one's pigs and cows.

Another big leap forward in mathematics germane to our present topic, occurred in the early 1800s at the hands of William Hamilton with the invention of "Quaternions" (the precursor to our present day vectors). Hamilton, being somewhat obsessed with the notion of grouping related numbers together, pressed to develop the concept, despite the extreme ridicule of his peers (who considered the concept an utter abomination, as well as a scourge to mathematics and mankind in general). Undaunted, Hamilton continued to tirelessly pursue his work until he created what became vector mathematics. This of course is rather fortunate for us, since without vectors, no major bridge, sky scraper, jet aircraft, laser, microwave, television, Cell phone (etc.) could be competently designed or understood today.

Obviously, other major developments have occurred in mathematics this past 150 to 200 years that have also come in handy, including Fourier Mechanics, Maxwell's Equations and Quantum Physics. Some of these seemingly peculiar mathematical tools in fact have proven exceptionally useful through the years, despite their apparent "imaginary" or "nonsensical" nature at the time. These tools have proven not only incredibly *powerful* in the hands of those trained in their use, but indeed very *real*. (Which, actually is the point of this whole discussion.) Case in point, is the use of "Complex numbers".

A Complex number is simply a pairing of two numbers, much the same way a vector is a triplet or quadruplet of numbers, which share some related connections (in fact Hamilton in the 1830s helped codify the notion of complex numbers as well as vectors, by showing both were simply examples of grouping related numbers together to form vectors). For dubious historical reasons (few care to disturb), the two numbers in this "complex" pair are referred to as the "Real" part and the "Imaginary" part, represented by something similar to:

$$C = A + iB$$

Here, “A” represents the "real" part, and “B” represents the "imaginary" part of our complex number. Visually, these two vector components can be drawn in a 2-dimensional plane as shown in Figure D.1 by defining the horizontal axis as the “real” component, and the vertical axis to the “imaginary” component (note that this is merely to assist the user in visualizing the pairing of these two numbers, it is not necessarily representative of any physical structure or arrangement):

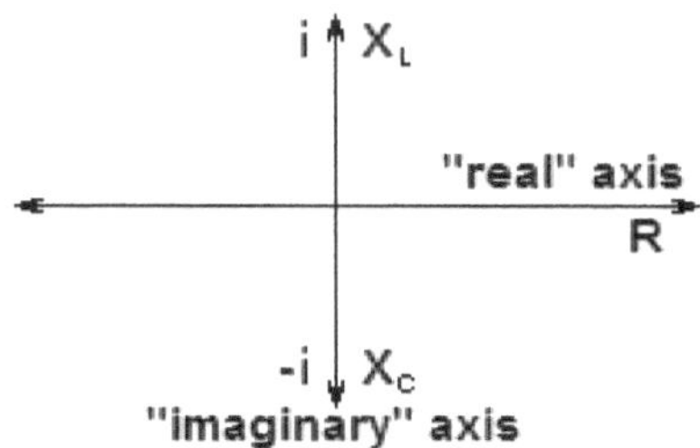

Figure D.1 Complex number plane.

By the late 1700s, mathematicians found that “imaginary” numbers enabled them to represent things that “ordinary” numbers could not encompass, including roots to negative numbers, as well as roots to “n^{th}” degree polynomials, etc. Though the unfortunate naming convention used in complex notation conveys something of a Lewis Carol, or Rod Serling connotation (e.g. "Real" and "Imaginary" numbers – second cousins to the mad hatter and magic-mushroom-dispensing caterpillars), the components of complex numbers are not really that "complex", nor are they all that "imaginary". Complex numbers allow us to represent things that are related, but orthogonal (e.g. Cos(x) and Sin(x), resistance and capacitive/inductive reactances, etc.) in a simple, yet elegant way.

Leonhard Euler in the last decade of the 1700s was the first to notice that the orthogonal functions of Cosine(x) and Sine(x) could be related to the natural exponential to help solve various differential equations. To do this, he simply paired them together using the “imaginary” number SQRT(-1), coined by

Euler as "i", with the identity which now bears his name (note that Engineering books often use "j" to represent SQRT(-1), in order to avoid confusion with current in a circuit):

$$\exp(ix) = \cos(x) + i\sin(x) \qquad \text{Euler relation} \quad (\text{D.1})$$

If we follow the same axis convention used in Figure D.1, we can plot Euler's relation in a way that expresses the orthogonality and interplay between these two functions in a particularly elegant and simple way. Since the maximum value of either the Cos() or the Sin() function is ±1, and since these two functions are 90° out of phase and orthogonal to one another (i.e. one is at its maximum value when the other is at zero, while 90° later the roles reverse), when we plot Euler's relation with Cosine(x) along the horizontal axis, and iSin(x) function along the vertical axis (with "x" being an angle from 0° to 360°), we see that Euler's relation plots a circle. Though Sin() and Cos() are the mathematical equivalent of oil and water, by drawing them along these perpendicular axes using "i" to accommodate their orthogonality, we are able to plot these two otherwise disjointed functions together on a single plot as a unit circle "bounded" by ±1:

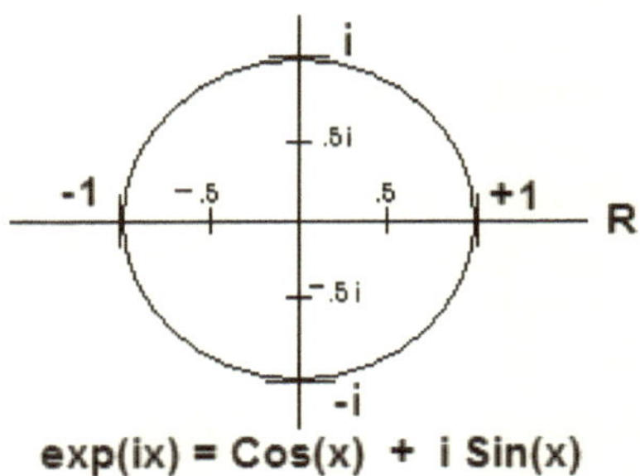

Figure D.2 Euler's relation and the unit circle.

With Euler's relation and this two-dimensional arrangement of "Real" and "Imaginary" axis, "i" becomes the equivalent of a 90° rotation. In other words, if we start with a "real" number and multiply it by "i", we rotate 90°. If we multiply that result by "i" again, we rotate it another 90°, or a total of 180°, putting the result back on the "Real" axis again, but now it is inverted (i.e. negative). Another multiplication by "i" again, and we are now inverted along the "imaginary" axis. And finally, one last application of an "i" and we return to our original starting point along the "real" axis. From this little exercise, we see that:

$$i * i = -1 \qquad \text{i.e.} \qquad i = \sqrt{(-1)}$$

A typical use of complex numbers common to most engineers and technicians, is the impedance (Z) in a circuit (a complex combination of resistance "R" and capacitance and/or inductive reactances "X", giving us the total impedance Z = R + iX). We find that though "R" and "X" limit current, we cannot simply add "R + X" to find the total impedance in a circuit (anymore than we would add "5 blocks north and 5 blocks east" to get "7 blocks as the crow flies"). Another engineering application of complex numbers is the cyclic phase nature of a propagating Electro-Magnetic wave ($E = Eo \exp[i\, \omega\, t]$), etc.

As for the terminology, we note that there is nothing "imaginary" about the effects of these quantities we draw along the "i" axis (for example, capacitive reactance X_C or inductive reactance X_L in a circuit). As most electronics buffs know, both capacitive and inductive reactances in a circuit have very *real* impact on current, indicating both are very *real* quantities. True, it is not as easy to measure X_C and X_L reactance with an Ohm meter or light bulb, they do nonetheless have a very significant impact on how the average electronic circuit operates. Reactance after all is why our audio signal "rolls-off" at the higher frequencies, why RF circuits oscillate, and why our digital pulse train "smears" as it travels through a system (i.e. "electronic dispersion"). Not only are these effects therefore very real, they require a *great deal* of effort (and *expense*) on our part to account for them.

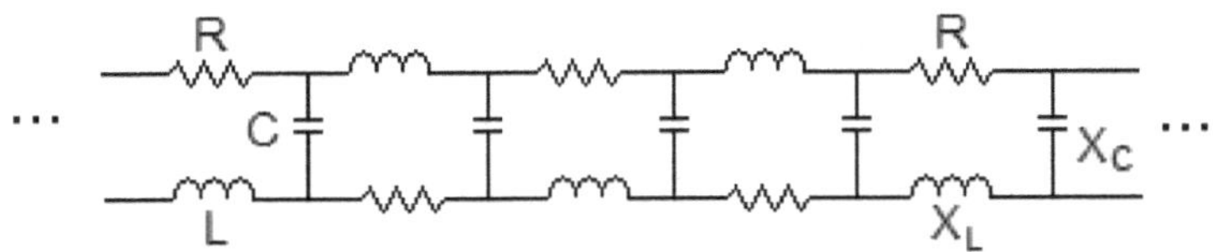

Figure D.3 Schematic representation of a simple coax line.

Appendix E: Dirac Notation

During the early years of Quantum Physics, two distinctly different notations emerged, both of which persist in present day literature. The first of these, "Heisenberg notation" is perhaps the most common for introductory discussions, since it writes everything in terms of differential equations. This approach allows the user to make a more tangible and direct correlation to familiar Classical Physics equations (such as our E&M Wave Equation, etc.). The second form is known as "Dirac Notation", (or occasionally as "Bra-Ket" notation), depicting all state functions as vectors and all operators as matrices.

As it turns out, the Dirac approach is a much more flexible and powerful form, allowing us to utilize the power and versatility of matrix manipulations in a slightly more abstract way, without necessarily requiring us to grind through the painful tedium of multivariable integration. To show the power of this approach, we will perform a few simple Quantum operations similar to those previously demonstrated, using the matrix form of Dirac notation.

To begin, we first define our state function as a superposition of various "$\Phi_{n(x)}$" components (as we did in Chapter 3):

$$\Psi_{(r)} = \sum a_n \Phi_{n(r)}$$

$$= a_0 \Phi_{0(r)} + a_1 \Phi_{1(r)} + a_2 \Phi_{2(r)} + a_3 \Phi_{3(x)} + \ldots$$

In Dirac / Bra-Ket notation, this is written as:

$$|\Psi_{(r)}> = \sum a_n |\Phi_{n(r)}>$$

We point out that our state function "$\Psi_{(r)}$" has not changed. It is still composed of a collection of Φ_n's as before. The only change is in the notation that allows us to write the state function as a collection of components in an array, much the same way you would write the components of a simple position or velocity vector as an array. In this "Bra-Ket" notation, the "Bra" form of our state vector becomes:

$$< \Psi_{(x)} | \quad = \quad [\, a_0\, \Phi_{0(x)}\,, \quad a_1\, \Phi_{1(x)}\,, \quad a_2\, \Phi_{2(x)}\,]$$

While the "Ket" form of our state vector is simply its transpose (see Appendix I for a review of matrix math):

$$| \Psi_{(x)} > = \begin{matrix} a_0\, \Phi_{0(x)} \\ a_1\, \Phi_{1(x)} \\ a_2\, \Phi_{2(x)} \end{matrix}$$

An operator (such as the Hamiltonian operator, "H" introduced in Chapter 3) is itself written as an array of discrete element:

$$\mathbf{H}_{op} \quad = \quad \begin{matrix} h_{11} & h_{12} & h_{13} & \ldots \\ h_{21} & h_{22} & h_{23} & \ldots \\ h_{31} & h_{32} & h_{33} & \ldots \\ & \ldots & & \end{matrix}$$

Therefore, when we apply a given operator on a specific state function, we simply perform a matrix multiplication of that operator against the state function.

<u>Example</u>: Energy Eigenvalue and Schrodinger's equation in Dirac notation:

Process Schrodinger's equation using the following simple Hamiltonian energy operator and state function:

$$\mathbf{H}_{simple} \quad = \quad \begin{matrix} \hbar & 0 & 0 \\ 0 & 2\,\hbar & 0 \\ 0 & 0 & \hbar \end{matrix}$$

The "Ket" form of the given state function is:

$$| \Psi_{(x)} > = \begin{matrix} \omega_0\, \Phi_{0(x)} \\ \omega_1\, \Phi_{1(x)} \\ \omega_2\, \Phi_{2(x)} \end{matrix}$$

Therefore, applying our "H" operator on this "Ket" state vector, we generate its Eigenvalue Schrodinger's equation:

$$H \; |\Psi_{(x)} > \quad = \quad E_n \; |\Psi_{(x)} >$$

Or:

$$\begin{bmatrix} \hbar & 0 & 0 \\ 0 & 2\hbar & 0 \\ 0 & 0 & \hbar \end{bmatrix} * \begin{bmatrix} \omega_0 \; \Phi_{0(x)} \\ \omega_1 \; \Phi_{1(x)} \\ \omega_2 \; \Phi_{2(x)} \end{bmatrix} = E_n \; |\Psi(x) >$$

Performing the indicated matrix multiplication (see Appendix I), we find the energy eigenvalue is:

$$H \; |\Psi_{(x)}> \; = \; \hbar \, \omega_0 \, \Phi_{0(x)} + 2\,\hbar \, \omega_1 \, \Phi_{1(x)} + \hbar \, \omega_2 \, \Phi_{2(x)}$$

Example Expectation value of position:

Use the same state function and the following operator to find the expectation value " $< \Psi| \; x \; | \; \Psi >$" of position (x):

$$\mathbf{R}_{\text{simple}} = \begin{bmatrix} a & 0 & 0 \\ 0 & b & 0 \\ 0 & 0 & c \end{bmatrix}$$

Combining our state function with this operator in "Bra-Ket" notation: $< \Psi_{(x)}| \; \mathbf{R} \; | \; \Psi_{(x)} >$

$$= \begin{bmatrix} \omega_0\Phi_{0(x)}, & \omega_1 \; \Phi_{1(x)}, & \omega_2 \; \Phi_{2(x)} \end{bmatrix} \begin{bmatrix} a & 0 & 0 \\ 0 & b & 0 \\ 0 & 0 & c \end{bmatrix} \begin{bmatrix} \omega_0 \; \Phi_{0(x)} \\ \omega_1 \; \Phi_{1(x)} \\ \omega_2 \; \Phi_{2(x)} \end{bmatrix}$$

$$= \begin{bmatrix} a\omega_0\Phi_{0(x)} & 0 & 0 \\ 0 & b\omega_1\Phi_{1(x)} & 0 \\ 0 & 0 & c\,\omega_2\Phi_{2(x)} \end{bmatrix} \begin{bmatrix} \omega_0\Phi_{0(x)} \\ \omega_1\Phi_{1(x)} \\ \omega_2\Phi_{2(x)} \end{bmatrix}$$

$$= \begin{bmatrix} a\,(\omega_0\Phi_{0(x)})^2 & 0 & 0 \\ 0 & b(\omega_1\Phi_{1(x)})^2 & 0 \\ 0 & 0 & c(\omega_2\Phi_{2(x)})^2 \end{bmatrix}$$

A similar approach can be applied to Schrodinger's Equation for the isolated Harmonic Oscillator[55] :

$$(P^2/2m + \tfrac{1}{2} k X^2)\ |\boldsymbol{\Psi}_{(x)}\rangle = \hbar\omega\,(v_{v00} + v_{0v0} + v_{00v} + {}^{3}/_{2})\ |\boldsymbol{\Psi}_{(x)}\rangle$$

and for a Harmonic Oscillator in an external Electric field (**E**):

$$(P^2/2m + \tfrac{1}{2} k X^2 - q\mathbf{E}X)\ |\boldsymbol{\Psi}_{(x)}\rangle$$
$$= \hbar\omega\,(v_{v00} + v_{0v0} + v_{00v} + {}^{3}/_{2}) + (q^2\mathbf{E}^2 / 2m\omega^2)\ |\boldsymbol{\Psi}_{(x)}\rangle$$

etc.

The main advantage of the Dirac notation, is that it allows us to easily depict our state function as vectors, and our operators as matrixes. This format allows us to take advantage of the "axioms" of matrix algebra to bypass the need to grind out the gory details of any "long hand" operations in the related differential equations.

---

55 Since we used orthogonal basis functions, we can use Separation of Variables to break Schrodinger's Equation for the Harmonic oscillator into several separate equations (one for each mode), and then combine them (or their individual energy eigenvalues) into a total composite when we are done (see Appendix G):

$$|\Psi_v\rangle = |\Phi_{v00}\rangle \ \mathbf{x}\ |\Phi_{0v0}\rangle \ \mathbf{x}\ |\Phi_{00v}\rangle$$

Appendix F: Hydrogen Atom wavefunctions:

The next two appendices are intended to provide a couple of concrete examples of the usefulness of the Quantum Mechanical tools developed thus far, by applying these tools in the quest to develop a good model of the Hydrogen atom, and later on the vibrational modes of molecules.

Here we will demonstrate more completely the Quantum model of the atom, presenting the following solutions to the Schrödinger equation for the Hydrogen atom. Hydrogen, being the simplest element in the Periodic Table (i.e. only 1 proton and 1 electron), enables us to derive a closed solution describing its electronic "cloud". To do this, we will exploit the orthogonality of our coordinate system to allow us to separate each of the variables describing its distribution over space and time (e.g. x,y,z,t). Unfortunately, solutions for the more complex atoms are not so easy to generate, primarily due to the screening effect the lower shell electrons create for those above them, making the closed solution for the simple Hydrogen atom an extremely important conceptual "stepping stone".

In essence, we find that the best we can do for more complex atoms/molecules is to generate *approximate* solutions, often made by perturbing or piecing together elements/concepts developed from the simple Hydrogen atom solution. Consequently, studying the Hydrogen atom serves a dual purpose, that of offering an instructional example into the power of Quantum Mechanics and the Schrödinger equation in particular, as well as offering something of a "starting point" or set of "building blocks" that allow us to model other more complex Quantum systems (see Chapter 5 and Appendix G).

To begin, we will use the classical approach to defining a system which focuses on the total energy in the system (KE + U). Recall that the total energy of a Quantum system can be described as a [quantatized] eigenvalue of the Schrödinger equation:

$$\mathbf{H}\,\Psi(r) \quad = \quad E\,\Psi(r)$$

Where "$\Psi(r)$" is the state function of the system, that represents the energy state the atom currently occupies, and **H** is the Hamiltonian energy operator introduced in Chapter 4.

An "excited" atom is one in which its electronic configuration has received energy from an external source, placing the electron cloud around the atom into a higher energy state. The most significant component of the energy in the system is due to the electrostatic force related to the separation between the positively charged nucleus, and the negatively charged electron "cloud". Other "secondary" energy components of the system include the energy contained in the orbital "angular momentum" of the electron, the electron "spin", etc.

The Hamiltonian energy operator (**H**) in the Schrödinger equation therefore is composed of all of these energy components[56] . Since the first two ("H_n" + "H_{ang}") are the most significant, we will concentrate most of our focus on these.

We define "H_n" as the operator corresponding to the energy due to radial separation between the nucleus and the electron, and "H_{ang}" as the operator corresponding to the energy in the "angular momentum" of the electron:

$$\mathbf{H} \quad = \quad H_n + H_{ang}$$

With the State function defined as "Ψ_{nl}" (with subscripts referencing the Quantum numbers described in Chapter 4), the total energy of the system can thus be extracted from the State function by applying the operators that correspond to these two energy components, i.e. the electron-nucleus separation (H_n) and the energy due to the angular momentum of the electron in its "Orbit" (H_{ang}):

---

56 In addition to the orbital angular momentum, we could also include the minute effects due to the "spin" of the electron (i.e. "spin-up" or "spin-down" orientation). To create our "total" state function we combine the "H_n" and "H_{ang}" with the spin states, to form a tensor product "Ψ_{nlm}". However, since the "spin" constitutes a significantly smaller energy component than even the orbital angular momentum, we will ignore its effects for now with negligible impact on our results.

$$E_{nl} = <\Psi_{nl} | H_n | \Psi_{nl}> + <\Psi_{nl} | H_{ang} | \Psi_{nl}>$$

$$= E_n + E_{ang}$$

(Note that <| |> is "Dirac" notation, and implies integration over all space; see Appendix E). Those readers familiar with Quantum theory (or the concept of a mathematical perturbation), should realize that, since the second term is much smaller than E_n, it in fact constitutes a slight perturbation to the dominant first term, thus reemphasizing the fact that the orbital angular momentum component makes but a small contribution to the total energy that only "tweaks" the orbital energy term.

Since the bulk of the energy in the system lies in the "H_n" term, we will deal with it first. Following the treatment used in Chapter 4, we write this term as a sum of the kinetic energy of the electron "orbiting" the nucleus + the potential energy of the attraction between the nucleus and the electron. The kinetic energy can be written in terms of momentum "p" (KE = $p^2/2m$), which we stated in Chapter 4 can be expressed as an operator "P" of the form $-i \hbar \partial_r$. Hence, the "H_n" term becomes:

$$H_n = P_r^2 / 2\mu + V_{(r)} = -\hbar^2 / 2\mu \ \partial_r^2 + V_{(r)}$$

If we let "r" be the orbital radius, then the potential $V_{(r)}$ around the atom is:

$$V_{(r)} = -Z e^2 / (4 \pi \varepsilon r)$$

To account for the "fine" detail energy "perturbation" due to the angular momentum of the electron, we now work our way through the second term in our initial Hamiltonian operator, i.e. "H_{ang}" written in terms of the angular momentum:

$$H_{ang} = L^2 \ \hbar^2 / (2\mu r^2)$$

Here "L" is the angular momentum operator which we can write in terms of "x,y,z" Cartesian coordinates, or more appropriately, in terms of Spherical coordinates "r", "θ" and "Φ" (with "θ" and "Φ" accommodating the angular dependence of the spherical geometry of the atom). Since the angular momentum "L" of any object is related to a change in its angular displacement ("θ" and "Φ"), it makes more sense to use the Spherical coordinate system rather than say the familiar Cartesian coordinate system. As we do, we will also find we'll need to use the corresponding momentum operators "P_θ" and "P_Φ" which are specifically related to "θ" and "Φ":

$$L^2 = L_x^2 + L_y^2 + L_z^2$$

$$= P_\theta^2 + (1/\sin^2\theta)\ P_\Phi^2$$

$$= -\hbar\,[\ \partial_\theta^2 + (1/\tan\theta)\,\partial_\theta + (1/\sin^2\theta)\,\partial_\Phi^2\ ]$$

For completeness, we will define an operator for the "z" component of the orbital angular momentum to round-out our set of operators:

$$L_z = -i\,\hbar\,\partial_\Phi$$

Since the "H_n" term has no angular dependence in it and the "L" terms have no radial dependence to them, they are independent of each other. This implies two things: 1) that we can separate the radial components from the angular components ("divide and conquer"), and 2) it therefore does not matter which order we apply each operator, i.e. these operators "commute":

$$H\,L^2 - L^2\,H = 0$$

which we will denote using the following notation:

$$[\,H, L^2] = 0$$

Furthermore, we note that varying powers of the same operator also commute, hence:

$$[\,L, L^2] = 0 = [\,L^2, L_z\,]$$

similarly:

$$[\,H, L_z\,] \;=\; 0$$

We note that not all operators commute. For example, it *does* matter a great deal which order we apply the "X" and "P" operators against a state function (see Chapter 4), hence they are *not* commuting observables. (As it turns out, this commutation property is tandem to the problem of trying to measure or "observe" position and momentum simultaneously with zero uncertainty, and thus relates directly to the whole Heisenberg Uncertainty principle.)

The happy coincidence that the H and L operators above do commute, allows us to use them to form what turns out to be a very important set, known as a "*Complete Set of Commuting Observables*" (C.S.C.O.), for use in describing this atomic system. Therefore, for Hydrogen-like atoms (e.g. singly ionized Helium, doubly ionized Lithium, etc.), we can write the eigenvalues for each of these commuting observables as (letting "H_{ang}" represent L^2, and "H^z_{ang}" represent L_z):

$$H_n \,|\,\Psi > \;=\; E_n \,|\,\Psi > \;=\; -\,Z^2e^2 / (4\,\pi\,\varepsilon\, n^2 <r_B>) \;\;|\,\Psi >$$

$$H_{ang} \,|\,\Psi > \;=\; E_{ang} \,|\,\Psi > \;=\; \hbar^2 \;l\,(\,l + 1) \;\;|\,\Psi >$$

$$H^z_{ang} \,|\,\Psi > \;=\; E^z_{ang} \;|\,\Psi > \;=\; m\,\hbar \;\;|\,\Psi >$$

Where: $<r_B>$ is the "average" value, or more correctly, the "expectation" value for the radius of the electron's "orbit", commonly referred to as the "Bohr radius" $r_B \;=\; \varepsilon\, h^2 / (\pi\, \mu\, e^2\,)$. To find "$H_n$" (the energy related to the orbital radius),we must take the expectation value of the radius, since the electron is not a hard little sphere, but is instead a "region" of charge with no hard-limit boundary or edge. This "average" radial position or expectation value is:

$$< r >_{nl} = < \Psi_{nl} | r | \Psi_{nl} >$$

$$= {}_{-\infty}\int\int\int^{\infty} \Psi^*_{nl} \; r \; \Psi_{nl} \; dv$$

$$= r_B [3n^2 - l (l+1)] / 2Z$$

(where the integral is taken over the full volume of space, dv = dr,dθ,dΦ).

We note that the state function "Ψ(r)" can be specifically derived in only the simplest of cases (e.g. the Hydrogen atom) from the Schrödinger equation using separation of variables. This approach allows us to break the state function into its separate orthogonal components, the first representing the orbital "radial" position component of the electron-nucleus system, and the second the angular distribution of the electron around the nucleus:

$$\Psi_{nlm\,(r,\theta,\Phi)} = R_{nl\,(r)} \, Y_{lm\,(\theta,\Phi)}$$

where "$R_{nl(r)}$" represents the radial dependence function of the electron cloud (note no angular dependence "m"), and "$Y_{lm(\theta,\Phi)}$" represents "Spherical Harmonic" functions (mentioned in Chapter 6), which represent the angular dependence of the electron (note that they have no dependence on radius "r" and hence no shell number "n"). The variables and constants used in the above equations (for Hydrogen-like atoms) are defined as follows:

$Y_{l,m\,(\theta,\Phi)}$ = Spherical harmonics (see Arfken [3])

$$R_{n,l\,(r)} = -\{ (2Z / nr_B)^3 (n-l-1)! / 2n[(n + l)!]^3 \} * e^{-\rho/2} \rho^l L^{2l+1}{}_{n+l(\rho)}$$

where: $L^{2l+1}{}_{n+l(\rho)}$ = Laguerre polynomials (see Arfken §13.2)

n = the radial position/shell Quantum number

l = angular momentum Quantum number

m = spin component Quantum number

(Note that "n" and "l", are strictly integer values)

Z = the number of protons in the nucleus

μ = the reduced mass = $m_{nuk}\, m_e / (m_{nuk} + m_e)$

L = Total angular momentum of the electron

For completeness, we now show a few of the solutions for the Hydrogen-*like* State functions (with corresponding graphical representations in Figure F.1).

Defining "a" as the Bohr radius ("r_B"), we have the following:

$$\Psi_{nlm} = R_{n,l\,(r)}\ Y_{l,m\,(\theta,\Phi)}$$

$$R_{1,0(r)} = 2\,(Z/a)^{3/2}\ e^{-Zr/a}$$

$$R_{2,0(r)} = 2\,(Z/2a)^{3/2}\ e^{-Zr/2a}\ (2 - Zr/a)$$

$$R_{2,1(r)} = (3)^{-1/2}\ (Z/a)^{3/2}\ e^{-Zr/2a}\ (Zr/a)$$

$$R_{3,0(r)} = 2/3\ (Z/3a)^{3/2}\ e^{-Zr/3a}\,(3 - 2Zr/a + 2/9\,[Zr/a]^2)$$

etc.

$Y_{0,0\ (\theta,\Phi)}$ = SQRT(1 / 4π) = 's' subshell (constant sphere)

$Y_{1,0\,(\theta,\Phi)}$ = SQRT(3 / 4π) Cos θ = 'p' subshell ('figure-8')

$Y_{1,\pm 1\,(\theta,\Phi)}$ = -/+ SQRT(3 / 8π) Sin θ $e^{\pm i\pi}$

$Y_{2,0\,(\theta,\Phi)}$ = SQRT(5 / 16π) (3Cos²θ – 1)

etc.

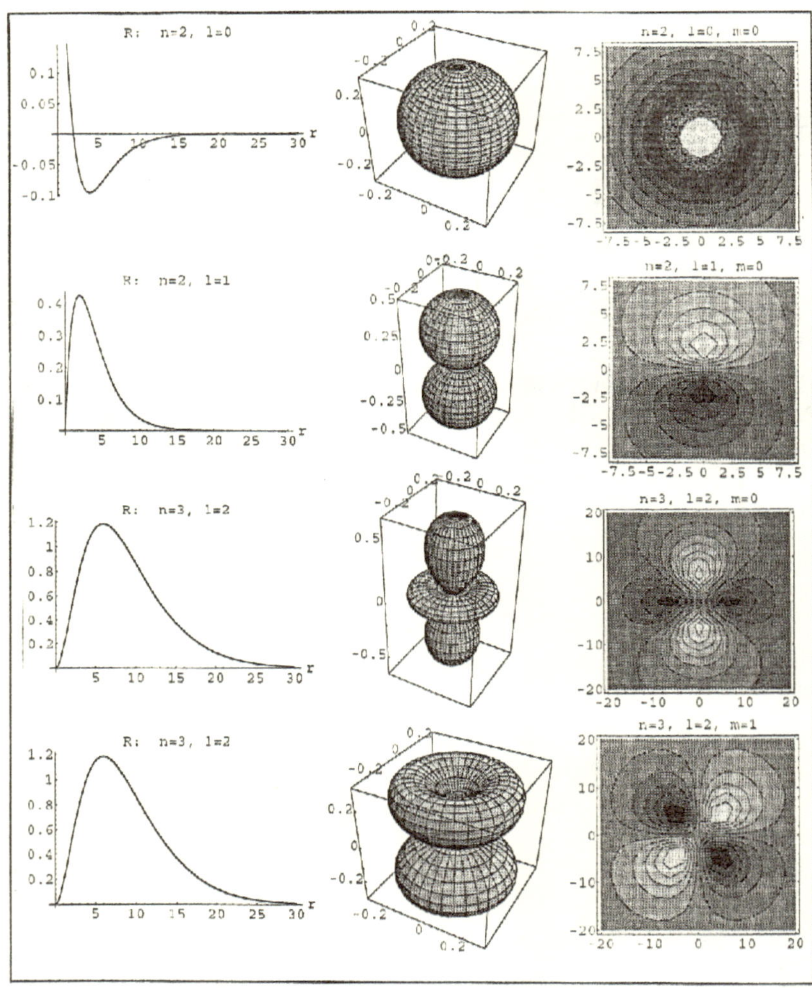

Figure F.1 Selected Hydrogen atom wavefunctions, $\Psi_{nlm\ (r,\theta,\Phi)}$ = $R_{nl\ (r)}$ $Y_{lm\ (\theta,\Phi)}$. The radial component is depicted down the left edge, and the angular dependence down the center. Grayscale key: white=+, gray=0, black= – .

As mentioned in the body of the text, the transitions between the various levels are not only restricted to integer changes, but are also governed by symmetry considerations (i.e. minimal overlap between levels). As a result, some orbital transitions have a high probability of occurring, while others have very low probabilities of occurring. These symmetry restrictions are specified by the "Dipole Selection Rules", as described in Appendix H. Since the Dipole Selection Rules tell us that the angular momentum "subshells" must change by one (i.e. "Δl" must = ± 1), we find that some atoms/molecules remain in an exited state for an extremely long period of time rather than transition immediately to the ground state, giving rise to what we have been calling the "metastable" states.

For example, we often encounter atoms which when excited into the "2s" state, take a very long time before they decay to the "1s state, which we would expect to occur quickly based solely on stability considerations (i.e. give off excess energy and drop to a more stable state). Again, the reason for this initially unexpected behavior is that the "2s" to "1s" transition is forbidden by this Dipole Selection Rules (since the subshell number "l" would not change in such a transition).

Note that in deriving these selection rules (see Appendix H), only dipole moments are considered. Higher moments, such as the quadrapole moment are assumed to be much smaller than dipole moments. However, even though they are much smaller than the dipole moments, they are nevertheless typically not zero. Consequently some very slight probability still exists for a "2s" to "1s" transition (which explains why this transition does eventually take place). This is indeed verified by spectral data, which indicates a very small percentage of "2s" metastable states do decay into any vacant "1s" states spontaneously. It is this spontaneous de-excitation that is responsible for the initial triggering of stimulated emissions.

Appendix G Schrödinger's equation and molecular vibrations:

Since most undergraduate level Chemistry and Physics courses tend to conclude their discussions on Schrödinger's equation without offering too many "real-world" applications beyond the Hydrogen atom, most students tend to come away from the discussion with the opinion that it's "interesting", but limited in its practical use (i.e. to the Hydrogen atom).

To counter such a misconception, we offer the following technique based in part on that used by Papoušek and Aliev [13]. This application demonstrates the versatility of Schrödinger's equation in its ability to solve a much wider range of problems than just the simple Hydrogen atom, including a solution for molecular vibration states. Since these vibrational modes generate observable spectra (in the Infrared region), we can verify the success of this effort in applying Schrödinger's equation by comparing its theoretical predictions with hard spectral data generated in the lab. In general, these results show remarkably good correlation between theory and fact, offering strong validation of both the general Quantum model, as well as Schrödinger's equation in particular.

In this application of Schrödinger's equation, we liberate ourselves from the obsession with the standard spatial coordinates (e.g. x,y, and z), and instead adopt a much more flexible perspective. Instead of defining Schrödinger's equation in terms of "x,y,z" or "r,θ,Φ", we will write it in terms of *generalized* "normal coordinates" "Q_k," (each orthogonal to the other, as is the case with "x,y,z" and "r,θ,Φ"). In this case, the Q_k normal coordinates are defined here to be the coordinates related to the "k^{th}" vibrational modes of the molecular "bond-spring" system. In this form, "k" is the index specifying the vibrational mode of the molecule, summing from the first vibrational mode, to its "N^{th}" vibrational mode (with each "k" mode capable of many energy levels; see Figure 5.9).

Using this approach, we will find that the Schrödinger Equation for the generalized molecular Harmonic Oscillator becomes:

$$(-\hbar^2/2\mu)\ \sum_k \partial_k^2\, \Psi_k^\omega \ + \ \tfrac{1}{2}\,\mu \sum_k \omega_k^2\, Q_k^2\, \Psi_k^\omega \ = \ E_k^\omega\, \Psi_k^\omega$$

Since "$\sum_k$" sums "k" from 1 to "N", this equation represents "N" Schrödinger equations (one for each vibrational mode). Here, "μ" represents the "reduced mass" (discussed in Chapter 5 and Appendix F), and "ω_k" specifies the fundamental frequency of oscillation of the kth mode. Hence the individual equation for the kth mode of vibration is:

$$(-\hbar^2/2\mu)\ \partial_k^2\, \Psi_k^\omega \ + \ \tfrac{1}{2}\,\mu\, \omega_k^2\, Q_k^2\, \Psi_k^\omega \ = \ E_k^\omega\, \Psi_k^\omega$$

(we will show the origins of this equation below).

Since Q_k represents displacement of the molecule from its resting position, we see that the second term in the above equations ($\tfrac{1}{2}\ \mu\ \omega_k^2\ Q_k^2$) represents the potential energy of a molecular Harmonic Oscillator (analogous to the potential energy stored in a classical spring, $\tfrac{1}{2}$ K X^2, where "K" is the bond "spring" strength factor, and is equal to "m ω^2"). As such, we know from our discussions in Chapter 2, that this spring-loaded oscillator has a resonant frequency. This resonant frequency is proportional to the strength of the restoring force in the "spring-like" bond of the molecule, and inversely proportional to the mass being moved by the spring (i.e. ω = SQRT[K/m]).

In general, a molecule has "N" different vibrational modes, each designated here by one of our "generalized coordinates", Q_k. In Chapter 5 for example, we found CO_2 had four vibrational modes (one of which is degenerate; see Figures 5.6 and 5.8). Each of these vibrational modes has "v_k" possible excitation levels. As a result, the energy stored in each vibrational mode for our harmonic oscillator can be written as a sum of all the possible excitation levels of that mode:

$$E_{k,v} \ = \ \hbar\omega_k\,(\, v_k + \tfrac{1}{2}\,) \qquad\qquad v_k = 0, 1, 2, \ldots$$

Since our "normal coordinates" Q_k's are defined to be orthogonal to one another (analogous to "x,y,z", "r,θ,Φ", etc.), we can use the same approach we used on the Hydrogen atom in Appendix F. In that case we used separation of variables on the "R,θ,Φ" state function to produce two separate equations: one related to the radial dependence ("R"), and one related to the angular dependence ("$Y_{\theta,\Phi}$"). Once solved, we then combined their solutions to create the overall state function for the Hydrogen atom.

Using separation of variables here on our molecular equation, we define "N" state functions for the molecule for each of the separate Q_k "spaces". The total state function of our system then is the tensor product of the separate state functions for each of the Q_k "spaces":

$$\Psi \quad = \Psi_{(Qk=1)} \otimes \Psi_{(Qk=2)} \otimes \Psi_{(Qk=3)} \otimes \ldots \Psi_{(Qk=N)}$$

Once we solve for the energy of each Q_k space, we then sum all these energies (the "v" index) for each separate vibrational mode Q_k (the "k" index) to get the total energy in our molecular oscillator:

$$E_{tot} \quad = \quad \sum_v E_{k=1,\,v} + \sum_v E_{k=2,\,v} + \sum_v E_{k=3,\,v} + \ldots$$

$$= \quad \sum_k \sum_v \hbar\, \omega_k \left(v_k + \tfrac{1}{2} \right)$$

To construct the Schrödinger's equations above, we use a Hamiltonian operator composed of the kth mode H_k's:

$$H \quad = \sum_k H_k \; = \; \sum_k (P^2_k / 2\,\mu) \; + \; U_{(r)\,k}$$

$$= \; (-\hbar^2/2\mu)\, \partial_k^2 \; + \; \tfrac{1}{2}\, \mu\, \omega_k^2\, Q_k^2$$

Therefore Schrödinger's equation ($H_k\, \Psi_k^\omega = E_k^\omega\, \Psi_k^\omega$) becomes:

$$(-\hbar^2/2\mu)\, \partial_k^2\; \Psi_k^\omega \; + \; \tfrac{1}{2}\, \mu\, \omega_k^2\, Q_k^2\; \Psi_k^\omega \; = \; E_k^\omega\, \Psi_k^\omega$$

The solution for a *three* mode molecular vibrator is:

$$\Psi_{(Q_1Q_2Q_3)} = \frac{\beta \exp(-\beta^2 [Q^2_{k=1} + Q^2_{k=2} + Q^2_{k=3}]) \, Ћ_{v_1(\beta)} \, Ћ_{v_2(\beta)} \, Ћ_{v_3(\beta)}}{(\pi \, 2^{(v_1+v_2+v_3)} \, v_{k=1}! \;\; v_{k=2}! \;\; v_{k=3}!)}$$

where $\beta^2 = \mu\omega/\hbar$, "$v_1$" is the vth vibration in the k=1 mode (etc.), and the "$Ћ_{v(\beta)}$" terms are Hermite Polynomials (see Arfken [3]).

Appendix H: Dipole Selection Rules:

As mentioned previously, the transitions between various energy levels in the atom are not only restricted to discrete orbital changes, but are also governed by symmetry considerations (i.e. minimal overlap between sub-orbitals in those energy levels). Consequently there is a high probability for some transitions to occur, while at the same time others have only very low probabilities. In Chapter 3 and Appendix F we mentioned that the sub-shell Quantum number "l" must change by *one*, i.e. $\Delta l = \pm 1$, and thus for example, a transition from a "2s" state to a "1s" state has a very low probability of occurring. This requirement stems from the "Dipole Selection Rules", which are based on the probability of overlap between two states, which is modeled in the equations below (assuming a dipole distribution of charge, i.e. a dipole defined as "**p** = e**r**"):

$$<|\mathbf{p}|> \; = \; < \Psi_{nlm} \,|\, e\,\mathbf{r} \,|\, \Psi_{nlm} >$$

$$= \iiint \Psi^*_{n'l'm'} \;\; e\,\mathbf{r} \;\; \Psi_{nlm} \; d^3r$$

To solve this overlap integral, we make the following definitions, using "Spherical Harmonic" functions to write the angular dependent functions, i.e.:

$$x = r \cos\Phi \sin\theta$$

$$= r/2 \;\sin\theta \,[\, e^{+i\Phi} - e^{-i\Phi} \,] \; = \; SQRT(2\pi/3) \;\, r \;[\, Y_{1,-1} - Y_{1,1} \,]$$

$$y = r \sin\Phi \sin\theta$$

$$= r/2 \;\sin\theta \,[\, e^{+i\Phi} + e^{-i\Phi} \,] \; = \; i\, SQRT(2\pi/3) \;\, r \;[\, Y_{1,-1} + Y_{1,1} \,]$$

$$z = r \cos\theta \; = \; SQRT(4\pi/3) \;\, r \;\, Y_{0,1}$$

Using the above identities involving Spherical Harmonics and writing our integral:

$$<|\mathbf{p}|> \;=\; \int\int\int \Psi^*_{n'l'm'} \; e\,\mathbf{r} \; \Psi_{nlm} \; d^3r$$

in spherical coordinates (with $d\Omega = \sin\theta \, d\theta \, d\Phi$), we end up with an integral with cross terms involving $Y_{1,-1}$, $Y_{1,1}$, etc. Since Spherical Harmonics form a set of orthogonal functions, i.e.

$$\int\int Y_{l',m'} \; Y_{l,m} \; d\Omega \qquad = \qquad \delta_{l',l} \, \delta_{m',m}$$

the orthogonality between them leads to our Dipole Selection Rules:

$$\Delta l = \pm 1, \qquad \text{and:}$$
$$\Delta m = \pm 1 \qquad \Delta m = 0$$

Thus the suborbital number ("l") must increase or decrease by one, while "m_{new}" must either equal "m_{old}" or change by 1.

As an example, we will consider the probability density ("$<\Psi| \; p \; |\Psi>$") of the "forbidden" transition $\Delta l=0$ from "2s" → "1s" under the dipole approximation, noting that "s" sub-orbitals correspond to "l=0" and "m=0", i.e. $Y_{0,0}$ (a constant sphere). Beginning with the "x" component ($x = r \cos\Phi \sin\theta$) of the dipole ("$\mathbf{p} = e\mathbf{r}$"):

$$<\mathbf{p}> \;=\; \int\int\int \Psi^*_{1,0,0} \;\; e\,\mathbf{r} \;\; \Psi_{2,0,0} \; d^3v$$
$$= \int\int\int R^*_{1,0(r)} \, R_{2,0(r)} \; e\,r\cos\Phi \, \sin\theta \;\; \sin\theta \; dr \; d\theta \, d\Phi$$

(where the extra "$\sin\theta$" next to "dr" is needed to relate the Cartesian coordinates to Spherical coordinates). If either the radial or the angular portion of this integral goes to zero, the whole integral goes to zero, implying that the transition from the "2,0,0" state to the "1,0,0" state would have a "zero" probability of occurring (under our assumption of a dipole charge distribution).

Focusing only on the angular portion of this integral for the moment, we have:

$$\int_0^{2\pi}\int_0^{\pi} \sin^2\theta \ \cos\Phi \ \ d\theta \, d\Phi = 0 \qquad (x)$$

Similarity for the "y" and "z" components we have:

$$\int_0^{2\pi}\int_0^{\pi} \sin^2\theta \ \sin\Phi \ \ d\theta \, d\Phi = 0 \qquad (y)$$

$$\int_0^{2\pi}\int_0^{\pi} \sin\theta \ \cos\theta \ \ d\theta \, d\Phi = 0 \qquad (z)$$

Since all three integrals evaluate to zero, we conclude that the probability of the "2s" to "1s" transition occurring (i.e. "l" transition from "l=0" in the second orbital to "l=0" in the first orbital) is *zero.*

Once again, we emphasize that in deriving these selection rules, we approximated the charge distribution around the atom as having a simple dipole moment ("p = e **r**"). In reality, we know that the distribution of charges around an atom do not form a perfect dipole configuration, but tend to have a wider distribution, i.e. tend to exhibit slight (but non-zero) multipole moments (e.g. quadrapole, octapole, etc.). As a result, some slight probability still exists for a "2s" to "1s" transition, depending on the distribution of charge unique to each species of atom, as can be verified by spectral data. The dependence on the unique electronic configuration explains why each type of atom has its own characteristic lifetime for its metastable states.

Appendix I: Matrix mathematics:

A matrix is in essence an ordered set of numbers or variables, much the same way a position or velocity vector is an ordered set of components:

Example 1, Position vector: **R** = (x, y, z)

Example 2, Velocity vector: **V** = (V_x, V_y, V_z)

Example 3, E&M Field tensor:

$$\mathbf{F}^{\alpha\beta} = \begin{bmatrix} 0 & -\mathbf{B}_X & -\mathbf{B}_Y & -\mathbf{B}_Z \\ \mathbf{B}_X & 0 & \mathbf{E}_Z & -\mathbf{E}_Y \\ \mathbf{B}_Y & -\mathbf{E}_Z & 0 & \mathbf{E}_X \\ \mathbf{B}_Z & \mathbf{E}_Y & -\mathbf{E}_X & 0 \end{bmatrix}$$

Grouping data/variables in a vector or matrix arrangement, not only makes the set more compact, but it allows us to manipulate these sets of components in a very efficient way (e.g. grouping the coefficients of several differential or simultaneous equations in a compact array, etc.). In a sense, matrix math is to a group of coefficients (for example), as algebra is to arithmetic. By learning and following the "shortcut" rules to manipulating these groups of numbers, we expedite and simplify the processing of the information the matrix represents. This is very analogous to our learning the rules of algebra which enabled us to "shortcut" repetitive addition and subtraction (e.g. 4+4+4+4+4, or 4*5 = 20).

In a similar way, we can define a matrix array as some set of components, or a set simultaneous equation, etc, with each having its specific meaning or identity associated with a specific position in the matrix array. As long as we faithfully maintain the relationship of each element in the array (defined by its position in the array), it offers us a compact and powerful means of manipulating/solving a variety of very complex problems in a relatively simple and elegant way. Since the position of data in each element corresponds to a specific meaning (e.g. x,y,z ordering in a vector), special care must be taken to insure that all the components of vectors and matrices be kept in their correct position/order relative to each other, or the vector/matrix manipulation breaks down.

In order to correctly manipulate these matrices as single entities while retaining the relationship of each element within the group, a set of matrix "rules" need to be followed that defines each operation. The definition of these "rules" are analogous to the rules and axioms used to define each operation used in general algebra, insuring that we arrive at the correct answer as we apply each "shortcut". The matrix processing rules are outlined below, followed by several application examples designed to illustrate the power of matrix manipulations:

Matrix definition:

A matrix is a set of related components or elements within a system which can be represented using a column/row ordering, with "N" columns, and "M" rows:

$$\mathbf{A}_{MN} = \begin{bmatrix} a_{11} & a_{12} & a_{13} & \cdots & a_{1N} \\ a_{21} & a_{22} & a_{23} & \cdots & a_{2N} \\ a_{31} & a_{32} & a_{33} & \cdots & a_{3N} \\ \vdots & \vdots & \vdots & \cdots & \vdots \\ a_{M1} & a_{M2} & a_{M3} & \cdots & a_{MN} \end{bmatrix}$$

The simplest matrix, is the identity matrix:

$$I = \begin{bmatrix} 1 & 0 & 0 & 0 \ldots & 0 \\ 0 & 1 & 0 & 0 \ldots & 0 \\ 0 & 0 & 1 & 0 \ldots & 0 \\ \vdots & \vdots & \vdots & \vdots & \vdots \\ 0 & 0 & 0 & 0 \ldots & 1 \end{bmatrix} \qquad \textbf{Eqn I.1}$$

Matrix Transpose:

To transpose a matrix, one simply swaps the rows with the columns:

$$[\mathbf{A}_{ij}]^T = \mathbf{A}_{ji}$$

$$\mathbf{A}^T = \begin{bmatrix} A_{11} & A_{12} & A_{13} \\ A_{21} & A_{22} & A_{23} \end{bmatrix}^T = \begin{bmatrix} A_{11} & A_{21} \\ A_{12} & A_{22} \\ A_{13} & A_{23} \end{bmatrix}$$

Matrix Addition:

The addition of two matrices is accomplished by adding like elements from each matrix together. Consequently, matrix addition is only defined when the two matrices being added have the same number of rows and columns:

$$\mathbf{A + B} \quad = \quad \Sigma_m \Sigma_n \ (A_{mn} + B_{mn})$$

Example I.1: Matrix addition:

$$\mathbf{A + B} = \begin{Bmatrix} 1 & 2 & 2 & 1 \\ 0 & 2 & 2 & 0 \\ 1 & 2 & 2 & 1 \end{Bmatrix} \begin{Bmatrix} 0 & 1 & 1 & 0 \\ 2 & 2 & 2 & 2 \\ 3 & 3 & 3 & 3 \end{Bmatrix} = \begin{Bmatrix} 1 & 3 & 3 & 1 \\ 2 & 4 & 4 & 2 \\ 4 & 5 & 5 & 4 \end{Bmatrix}$$

Matrix Multiplication:

When multiplying a matrix by a simple scale factor, we apply that factor to every element in the matrix:

$$\mathbf{B} = \mu_o \mathbf{A} = \begin{bmatrix} \mu_o A_{11} & \mu_o A_{12} & \mu_o A_{13} \\ \mu_o A_{21} & \mu_o A_{22} & \mu_o A_{23} \\ \mu_o A_{31} & \mu_o A_{32} & \mu_o A_{33} \end{bmatrix}$$

When multiplying two arrays together (e.g. a vector times a matrix, or a matrix times another matrix), the process is a bit more involved. In that case, the multiplication is performed by working row-column pairs.

Specifically, we multiply two arrays together by multiplying each element in the first ROW of the first array by the corresponding element in the first COLUMN of the second array, and summing the result of each product. That sum represents the first element in the product matrix. This process is then repeated by again taking the same first ROW of the first array and multiplying the corresponding elements in the SECOND COLUMN of the second array, and summing (which is the next element in the first row of the product matrix). This process is repeated against all columns in the second array. Following that, the SECOND row of the first array is applied against the FIRST column of the second array (which produces the first element in the next row of the product matrix), and so on through all the columns of the second array.

From this it should be clear that the "width" (i.e. the number of columns) of the first array must equal the "height" (i.e. the number of rows) of the second array.

$$\mathbf{C}_{m\,n} = \mathbf{A} * \mathbf{B} = \sum_m \sum_n (A_{mn} * B_{nm})$$

where "C's" dimensions are "m x n")

Example I.2: Matrix multiplication:

$$\mathbf{A} * \mathbf{B} = \begin{bmatrix} 1 & 1 & 1 & 0 \\ 0 & 2 & 2 & 2 \\ 1 & 2 & 3 & 3 \end{bmatrix} \begin{bmatrix} 1/3 \\ 1/3 \\ 1/3 \\ 1/3 \end{bmatrix} = \begin{bmatrix} 1/3 + 1/3 + 1/3 + 0 \\ 0 + 2/3 + 2/3 + 2/3 \\ 1/3 + 2/3 + 3/3 + 3/3 \end{bmatrix} = \begin{bmatrix} 1 \\ 2 \\ 3 \end{bmatrix}$$

<u>Matrix Division</u>:

Matrix division is a bit trickier than matrix multiplication, since we are not dividing by a single number, but rather a group of numbers. Instead of dividing matrix "A" by matrix "B", we must multiply the first matrix "A" times the *inverse* of the second "B":

$$\mathbf{C} = \mathbf{A} * \mathbf{B}^{-1}$$

Matrix Inverse:

A matrix multiplied by its inverse produces the identity matrix (Eqn I.1), analogous to multiplying any number by its inverse: $x * 1/x = 1$:

$$\mathbf{A} * \mathbf{A}^{-1} = \mathbf{I}$$

We note however that not all matrices have an inverse. Only a square, non-singular matrix has an inverse.

By way of practical application, we note that there are a number of methods that can be used to generate the inverse of a matrix (e.g. Gaussian elimination, elementary row manipulations, etc.). We conclude this section with the following algorithm of this author's preferred method, for those so inclined:

Step 1: Write down the NxN matrix to be inverted "M". Create a copy of this as matrix "A", and append to this an equal number of columns to its right (i.e. "A's" size is Nx2N, rows x columns). Set the added columns equal to the identity matrix.

Step 2: Create a temp matrix "B" equal in size to the original matrix "M". Create another temp matrix "C" equal in size to the expanded matrix "A" (i.e. Nx2N).

Step 3: Begin loop "k", where "k" is the column index that runs from 1 to N. Equate matrix "B" to the identity matrix (Eqn I.1).

Step 4: Set B's k^{th} column's diagonal element (i.e. its diagonal element in that column) $B_{K,K} = 1/A_{K,K}$.

Step 5: Set B's non-diagonal elements in the "k" column $B_{R,K} = -A_{R,K} / A_{K,K}$ (where "R" designates the Row index).

Step 6: Perform a matrix multiplication: $\mathbf{C} = \mathbf{B} \times \mathbf{A}$ across *all* columns of "A".

Step 7: Increment "k" (i.e. advance to the next column), looping back to step 3, using the result matrix "C" in place of matrix "A" (i.e. do not use any part of the old matrix "A" again).

Step 8: Continue looping through the columns (index "k") until "N" columns in "C" have been processed. The final version of "C" should equal an identity matrix on its left half, and the inverse of the original matrix on its right half (assuming "M" does have an inverse; recall that not all matrices do).

Step 9: Verify that the product of "M" and the inverse found in the previous step equals the identity matrix: $\mathbf{M} \times \mathbf{M}^{-1} = \mathbf{I}$

Example: Finding the inverse of a matrix.

Given a 3x3 matrix "M", find its inverse using the technique outlined above.

$$M = \begin{bmatrix} 2 & 1 & 2 \\ 1 & 3 & 1 \\ 4 & 1 & 5 \end{bmatrix}$$

$$A = \left[\begin{array}{ccc|ccc} 2 & 1 & 2 & 1 & 0 & 0 \\ 1 & 3 & 1 & 0 & 1 & 0 \\ 4 & 1 & 5 & 0 & 0 & 1 \end{array}\right]$$

$$B = \begin{bmatrix} 1/2 & 0 & 0 \\ -1/2 & 1 & 0 \\ -2 & 0 & 1 \end{bmatrix}$$

$$C = \left[\begin{array}{ccc|ccc} 1 & 1/2 & 1 & 1/2 & 0 & 0 \\ 0 & 5/2 & 0 & -1/2 & 1 & 0 \\ 0 & -1 & 1 & -2 & 0 & 1 \end{array}\right]$$

$$C = \left[\begin{array}{ccc|ccc} 1 & 0 & 0 & 2.8 & -0.6 & -1 \\ 0 & 1 & 0 & -1/5 & 2/5 & 0 \\ 0 & 0 & 1 & -2.2 & 0.4 & 1 \end{array}\right]$$

Bibliography

1. de Solla Price An ancient Greek Computer, Scientific American June 1969

2 Mumma, Buhl, Chin et al. Discovery of natural gain amplification in the 10-micrometer carbon dioxide laser bands on Mars - A natural laser Science, vol. 212, Apr. 3, 1981, p. 45.

3 G. Arfken. Mathematical Methods for Physicists, Academic Press.

4 J. D. Jackson Classical Electrodynamics John Wiley & Sons publishing, 1975.

5 Reitz, Milford and Christy Foundations of Electromagnetic Theory Addison Wesley, 1980.

6 Milloni and Eberly Lasers John Wiley & Sons publishing, 1988

7 Harris and Bertolucci Symmetry and Spectroscopy Dover Publications, 1989.

8 Born and Wolf Principles of Optics Cambridge press, 1999.

9 E. Kapon Semiconductor Lasers II: Materials and structures (optics and photonics) Academic Press, 1999.

10 A. Gerrard and J Burch Introduction to Matrix Methods in Optics. Dover Publications.

11 G Fowles Introduction to Modern Optics. Dover Books.

12 C. Cohen-Tannoudji, et al. Quantum Mechanics John Wiley & Sons publishing, 1977.

13 Papoušek, D. and M. Aliev "Studies in Physical and Theoretical Chemistry", V.17 Molecular vibrational-rotational spectra N.Y. Elsevier scientific publishing company.

14 Halliday and Resnick Fundamentals of Physics. John Wiley & Sons publishing, 1988.

15 F. Kapron et al, Journal of lightwave technology, Aug 1989.

16 J Verdeyen Laser Electronics. Prentice-Hall publishing.

17 Nature 344, 524-526 (1990). (Article on IBM STM manipulation of atoms).

18 T. Milster, W. Jiang et al, A single-mode high-power Vertical Cavity Surface Emitting Laser, Applied Physics Letters Vol 72, No. 26, 29 June 1998.

19 V. Gubsky, A. Skorucak et al, Fabricaiton of long-period fiber gratings with no harmonics IEEE Photonics Techologhy Letters, Vol II, No. 1 January 1999.

20 Barga, Hager and Kendrick Coil laser document number DE0104 USAF Research Lab, Kirtland AFB, NM, USA.

www.ingramcontent.com/pod-product-compliance
Lightning Source LLC
Chambersburg PA
CBHW031821170526
45157CB00001B/135

9781312967250